Seshat Anthology
Volume 5

Editors
Daivat Bhavsar
Austin Albert Mardon
Catherine Mardon
Ehimen Ogadu

Graphic Designers
Susie Woo
Amna Zia

Typeset
Susie Woo

List of Contributing Authors
Austin Albert Mardon
Cheng En Xi
Rosalind Fleischer-Brown
Jasrita Singh
Aleefa Devji
Parmpreet Kang
Ananda Majumdar
Patricia D'souza
Fatima Saleem
Jessica Henschel

First Printing: 2021
Cover Design by Susie Woo and Amna Zia
Typeset by Susie Woo
ISBN 978-1-77369-224-1
Golden Meteorite Press
103 11919 82 St NW
Edmonton, AB T5B 2W3
www.goldenmeteoritepress.com

Seshat Anthology
Volume 5

Austin Albert Mardon, Catherine Mardon, Daivat Bhavsar, Ehimen Ogadu, Susie Woo, Amna Zia

Antarctic Institute of Canada

Featured Article Themes
COVID-19: Immigration, Ethnic Communities, Mental Health & Vaccine
Fetal Alcohol Syndrome Disorder
Bacterial Specificity
Homesickness
Dendrochronology

Financial Acknowledgement
The publication of the Seshat Anthology Volume 5 is financially supported by the Golden Meteorite Press, Antarctic Institute of Canada.

Contents

Children of the Migrant Community and COVID-19 Scenario

Ananda Majumdar, Austin A Mardon

Ananda Majumdar

The University of Alberta (Bachelor of Education after Degree Elementary, Faculty of Education, Community Service-Learning Certificate and Certificate in International Learning, CIL), Intern, Museum of Diaspora, Migration, GRFDT, New Delhi, India, Book Pecker Fellow, Peace X, India

Austin A Mardon

CM, KCSS, FRSC, FRCGS
Assistant Adjunct Professor, Department of Psychiatry & John Dossetor Health Ethics Centre, University of Alberta; Special Advisor, Glenrose Rehabilitation Hospital

COVID-19 builds the world vulnerable. It is a mass global disruption. People lost their jobs and have been economically insecure. ⁱIn South-East Asia and the Pacific, an estimated 11.6 million people are migrants, estimated 5.2 million are women out of 11.6. In this region of Asia, migrants are the resources for economic production and functions. Isolation and restriction over mobility increased the risk of discrimination, mistreatment, especially on women migrant workers and children. Educational institutions, schools have been closed due to this disaster and migrant children have grimly vulnerable. School is not an educational institution but a heaven for those displaced children as a source of food, an opportunity to identify abuse, and a centre for receiving the news. Due to the closure of the school and other educational institutions, they have lost their safety devices, which are a part of child protection services. This is how migrant and displaced children lost access to educational materials and faced an obstacle in accessing online learning occasions. The approach over migrants

1

(women and children) was not good ever, which is now in its worst image towards those people of every country. [ii]In ASEAN countries, reports have come that verbal abuse has been increased over certain nationalities. Citizens of ASEAN countries are now believing that migrants are the main source of these pandemics. The image of disgrace and exploitation are not common but dangerous for any society and section. [iii]This is how fanatical and gender-based violence is increasing in every country in the context of the COVID-19 scenario. women [iv]migrant workers are working in food processing, hospitals, caregivers for seniors, seafood processing, frontline worker etc. they have a greater risk of transmission of the disease due to their working environment and conditions. The conditions of the working environment for migrant workers are never healthy anywhere but they work because of their financial conditions and various social and cultural limitation. They do not receive jobs anywhere or everywhere due to identity. [v]Migrant without documentation is fearing for deportation, neither they are receiving health protection information, nor they are included with nationals for public health safety package. In some occupations, they are receiving less wages than others resulting from the creation of a wage-gap. This is how the existence of discrimination shows-up. [vi]In this situation, States have been called by the UNICEF to take measures for the protection of all kinds of migrant's community their women and children. They should develop measures for the access to essential services for all, communicate with migrant's community and campaign to stop racism and hate to those minor section of every country. Migrant and displaced children are at risk of missing out on accurate public health information, which is due to language barriers, undocumented situations. undocumented children are fearing to contact public authorities. Fake information about the spread of COVID-19 aggravated racism and nationalism against migrants' children who are in various states in their destination countries from their original countries. Sudden restrictions in travel, and on other things weakens children's safety worldwide. The food distribution process has been disrupting in the refugee camps and shelters. [vii]Countries like Yemen where an estimated 1/3 of children are so skinny and an estimated 80% of the population depends on food assistance are in disrupted condition; relief activities have been reducing due to air travel restrictions. It is a serious situation in Syria, Yemen. UN authorities are suspending their work on resettlement programs in many countries due to COVID-19, this is how an uncertain environment especially for the migrant communities and

2

their children has created. [viii]On April 22nd, 2020 estimated 57 countries temporarily suspend their air travel globally. [ix]They also decided on no exception for asylum seekers. [x]In the United States, asylum seekers including their children have been deported to Mexico or the Southern border as a part of their COVID-19 response. Lockdowns and quarantines have influenced many people in the world. [xi]In Ethiopia estimated 3272 returning people have been registered and quarantined at many places that include 434 lonely children. Unfortunately, a major portion of the population did not go for health screening, an estimated 135 were girl's child. UNCHR has called to every agency, country, and policymakers for the respecting international human rights and refugee protection standards. According to the UN, report children have faced problems in four areas due to pandemics.

References

1. You, A., Lindt, N., Allen, R., Hansen, C., Beise, J., & Blume, S. (2020). Migrant and displaced children in the age of COVID-19: How the pandemic is impacting them and what can we do to help. Migration Policy Practice, 10(2), 32-39. Retrieved from https://www.unicef.org/media/68761/file.

2. Douglas, J., Hulshof, K., Motus, N., Naciri, M., Nishimoto, T., & Blume, S. (2020). End stigma and discrimination against migrant workers and their children during the COVID-19 pandemic. UNICEF-East Asia and Pacific. Retrieved from https://www.unicef.org/eap/press-releases/end-stigma-and-discrimination-against-migrant-workers-and-their-children-during.

Endnote

[i]End stigma and discrimination against migrant workers and their children during the COVID-19 pandemic. Paragraph 1st.

[ii]End stigma and discrimination against migrant workers and their children during the COVID-19 pandemic. Paragraph 5th.

[iii]End stigma and discrimination against migrant workers and their children during the COVID-19 pandemic. Paragraph 5th.

[iv]End stigma and discrimination against migrant workers and their children during the COVID-19 pandemic. Paragraph 6th.

[v]End stigma and discrimination against migrant workers and their children during the COVID-19 pandemic. Paragraph 7th.

[vi]End stigma and discrimination against migrant workers and their children during the COVID-19 pandemic. Paragraph 8th.

[vii]Migrant and displaced children in the age of COVID-19: How the pandemic is impacting them and what can we do to help. Legal shift. Paragraph 1st.

[viii]Migrant and displaced children in the age of COVID-19: How the pandemic is impacting them and what can we do to help. Legal shift. Paragraph 2nd.

[ix]Migrant and displaced children in the age of COVID-19: How the pandemic is impacting them and what can we do to help. Legal shift. Paragraph 2nd.

[x]Migrant and displaced children in the age of COVID-19: How the pandemic is impacting them and what can we do to help. Legal shift. Paragraph 2nd.

[xi]Migrant and displaced children in the age of COVID-19: How the pandemic is impacting them and what can we do to help. Legal shift. Paragraph 4th.

**In India there is a second wave of COVID-19 which is worse than the first wave of pandemics. Over a billion people in India are facing a tremendous challenge to tackle this global disease currently. Millions of people died and been affected already. The government is facing challenges to control the disease in India and the subcontinent.

"COVID—19 and Immigration Portfolio"

Ananda Majumdar

Ananda Majumdar

The University of Alberta (Bachelor of Education after Degree Elementary, Faculty of Education*
Community Service-Learning Certificate and Certificate in International Learning, CIL) *

Harvard Graduate School of Education (Professional Education as a Child Development Educator, Certificate in Early Education Leadership (CEEL-Series 2), online) *

Prospective Summer School in History and Archeology (2021-22)- "On the footsteps of Jesus: Jerusalem to Magdala, European University of Rome, Italy

Certificate in Migration Studies, GRFDT, New Delhi, India (Online)

Grant MacEwan University (Diploma in HR Management)

Jadavpur University (Master of Arts in International Relations) Sikkim Manipal University (Master of Business Administration in HR and Marketing Management)

MBB College, Tripura University (Bachelor of Arts in Political Science) Antarctic Institute of Canada (Researcher and Writer), Servicing Community Internship Program (SCiP) Funded by the Government of Alberta Member of Student Panel, Cambridge University Press, Member of the Association of Political Theory (ATP) University of Massachusetts

Student Member of ESA (European Studies Association), Columbia University, U.S.

General Coordinator, Let's Talk Science, University of Alberta

Early Childhood Educator, Brander Garden After School Parents Association

Book Pecker Fellow, Peace X, India

Intern, Migration, GRFDT, New Delhi, India

Cell# 1-780-660-7686

Edmonton, Alberta, Canada

anandamajumdar2@gmail.com, ananda@ualberta.ca

Global migration is not a new term. People are migrating through various ways one to other countries from ancient times. Current coronavirus has changed the mindset of global leaders. They are now taking initiatives to control migration in their countries by closing borders for a long time to permanent. The effect of postponing migration has influenced the individual economy and countries economy. In the long run, this restriction would continue for the fear of the invisible virus and the protection of human beings. The virus is coming through stages such as first wave, second wave, third wave etc. therefore everyone is afraid about its outcome. Political will force is not seeking for the solution of migrant community problems through structural equality. They are now more vulnerable due to several restrictions such as moving restrictions implementation by the global leaders. The leaders do not want migrant people and their global movement for work and other activities. Nationalism rises everywhere, therefore the importance of the citizens has been prioritized rather than the minority community and their other sections. The most vulnerable section under the migrant section is undocumented workers. They are facing problems mostly due to no identity. They do not access government-provided health checks and other facilities. COVID-19 has prejudiced every country in the world, resulting in estimated [i]10 million cases and an estimated 500 000 deaths worldwide. It has spread among vulnerable

6

communities in every republic such as ethnic minorities who are the victim of social problems such as low income, less socio and economic status etc. Migrants who are lawfully residing are facing various injustices in the field of economic and health sectors. Undocumented migrants are in fear of deportation at any time. It is almost impossible in a general pathway (instead of the firewall) for them to get access to a benefit such as in health care currently. The current pandemic makes them helpless due to their identity as the hidden community. In the United States, the ratio of the affected Afro-Americans and the Hispanic community is higher than the White population. An estimated 34% affected by the Corona Virus within the total population of the United States and they are estimated to only 18% of the entire population. In the United Kingdom, Asian and Black minorities are dying more than the White population. Migrants living under refugee camps, detention centres are at high risk of COVID-19 revelation. Therefore, where should they go? Who will help them? UNICEF works worldwide to protect the rights of minorities and migrant people. UNICEF is also working for displaced children by providing life protection equipment in refugee camps, child-friendly spaces for safe play. UNICEF and its other nongovernmental organization (NGOs) are working with the government, the private sector, and the civil societies of every country for the solution of migration, minority problems. The Global Refugee Compact is an international agreement that has set up several building blocks for a stronger and positive global response on refugee issues. In 2018 GRC provides a roadmap to various humanitarian international organizations, and host countries for the betterment of minority and migrant communities. They urge them for the integration of the migration community into the mainstream of society. [ii]GRC implements four objectives for the development of minorities and migrants. GRC keeps pressures on host countries, increases refugee self-reliance, expansion of resettlement and its access by those community and a safe return of all refugees, migrant temporary workers, or any minority class to their original countries. UNICEF and UNCHR are committing Global Compact on Refugees and make a blueprint for joint action with UNCHR for the reaching goals of migrant's problems solution. The blueprint of the UNICEF with the collaboration of UNCHR is a good document for a better practice around the world to support the children and youth of the refugees, minorities, and migrants. The Global Compact for the safe and regular pathways of migration is a milestone that has recognized the first time that children are central to migration management. UNICEF has worked a lot for the finalization of the document to protect the rights of the children of deprived communities in every country around the world and the children of the migrant community. On the other side host countries of the migrant community can take several steps for the betterment of the migrant community. They can make sure to provide access to health care, basic utensils, and services for livelihood. [iii]Portugal has provided temporary citizenship rights to all its asylum seekers and other migrants. [iv]Italy has declared for a temporary work permit to its all (estimated more than half-million) undocumented migrants. California State has declared to provide $500 to all undocumented workers for their feed and other necessary utensils. It is necessary for access to rights, access to health care, access to education, access to free movement and mobilization, access to various state benefits that other regular citizens receive. It is a key for every undocumented people, refugees, under-documented people as well for their livelihood in daily life. Without documentation, they can not access or received it difficultly under a firewall, temporarily such as the situation of the pandemic period. But in a normal situation (regular pathway) they are unable to get it. Life is surrounded by various dangerous things such as fall into the trap, criminal activities. They need to come out of this criminal activity. Therefore, they need national identity through a birth certificate, passport etc. otherwise identifying, or classifying is a challenge for authority. Documentation is a pathway for a normal life with mobilization including all kinds of benefits and rights of the human being. Otherwise legally migrants can not claim anything in the fear of deportation. The scenarios of global migration were never bad. It was a symbol of multiculturalism, bilingualism through which a country could engage with diverse people from around the world. COVID-19 pandemic has changed the image entirely. The post-COVID-19 environment could be different. Because of health-conscious, the term migration may be abolished by world governance for more health security. But developing countries like Bangladesh, the Philippines are highly dependent on migrant's remittance. Those countries financially may suffer along with the problem of the migrant family. Therefore, a policy should be made by the world leaders or by a legal organization like UNICEF for a better understanding.

End Note

[i]COVID-19: Exposing and addressing health disparities among ethnic minorities and migrants. Paragraph1

[ii]Migrant and displaced children. Global Refugee Compact. Paragraph 1

[iii]As Covid-19 rages on, countries need to support migrant workers. Solution. Paragraph1

[iv]As Covid-19 rages on, countries need to support migrant workers. Solution. Paragraph1

References

1. Least Protected, Most Affected: Migrants and refugees facing extraordinary risks during the COVID-19 pandemic. (2020). The New Humanitarian. Retrieved from October 09, 2020, from https://www.csis.org/analysis/five-ways-covid-19-changing-global-migration

2. Chugh, A. (2020). Will COVID-19 change how we think about migration and migrant workers? World Economic Forum. Retrieved from May 22, 2020, from

"COVID-19 and the Challenges of Ethnic Communities"

Ananda Majumdar

Ananda Majumdar

The University of Alberta (Bachelor of Education after Degree Elementary, Faculty of Education* Community Service-Learning Certificate and Certificate in International Learning, CIL) *

Harvard Graduate School of Education (Professional Education as a Child Development Educator, Certificate in Early Education Leadership (CEEL-Series 2), online) *

Prospective Summer School in History and Archeology (2021-22)- "On the footsteps of Jesus: Jerusalem to Magdala, European University of Rome, Italy

Certificate in Migration Studies, GRFDT, New Delhi, India (Online)

Grant MacEwan University (Diploma in HR Management)

Jadavpur University (Master of Arts in International Relations) Sikkim Manipal University (Master of Business Administration in HR and Marketing Management)

MBB College, Tripura University (Bachelor of Arts in Political Science) Antarctic Institute of Canada (Researcher and Writer), Servicing Community Internship Program (SCiP) Funded by the Government of Alberta Member of Student Panel, Cambridge University Press, Member of the Association of Political Theory (ATP) University of Massachusetts

Student Member of ESA (European Studies Association), Columbia University, U.S.

General Coordinator, Let's Talk Science, University of Alberta

Early Childhood Educator, Brander Garden After School Parents Association

Book Pecker Fellow, Peace X, India

Intern, Migration, GRFDT, New Delhi, India

Cell# 1-780-660-7686

Edmonton, Alberta, Canada

anandamajumdar2@gmail.com, ananda@ualberta.ca

Introduction: COVID-19 has prejudiced every country in the world, resulting in estimated [1]10 million cases and estimated [2]500000 death worldwide. It has spread among vulnerable communities in every republic such as ethnic minorities who are the victim of social problems such as low income, less socio and economic status etc. Migrants who are lawfully residing are facing various injustices in the field of economic and health sectors. Migrants who are undocumented are in fear of deportation at any time. It is almost impossible in a general pathway (instead of the firewall) for them to get access to a benefit such as in health care currently. The current pandemic makes them helpless due to their identity as the hidden community. In the United States, the ratio of the affected Afro-Americans and the Hispanic community is higher than the White population. An estimated 34% affected by the Corona Virus within the total population of the United States and they are estimated only 18% of the entire population. In the United Kingdom, Asian and Black minorities are dying more than the White population. Migrants living under refugee camps, detention centres are at high risk of COVID-19 revelation. The misfortune of vulnerable communities is in full swing for a long time. They are the victim of the hate mongers. Political will force is getting an extreme level to force against minorities everywhere in the world. They are wishing to postpone the migration process. Nationalism hates speech are dangerous narratives for the minorities of every country. The government of Canada has described elaborately the narratives of vulnerability and the communities. [3]First, they have defined vulnerability. They said vulnerable(weak) are older adults, people of any age along with chronic conditions, people with language problems, disabilities, people with social and geographic isolation(minority) and many other definitions of vulnerable groups. In these pandemic circumstances, the Canadian Government advises following them and their pandemic bulletin. The government said that initiative like information, data needs to be collected and collaborate with agencies, government for teamwork to reduce the virus. Plan for potential disruption, prepare for the shelter and shared space limitations and following all the cleaning rules. Therefore, Canada defines vulnerability in a different way, it is not only a minority group but the other characters of the human community as well. [4]United Nations defines vulnerable people through various narratives such as national, ethnic, religious and linguistics, geographical distance etc. United Nations makes sure that the group of those characteristics enjoy their rights through the practicing of their own culture and religion in any part of the world. They are a special group because of their identity as a minority section. Under the declaration on the rights (1992) of persons belonging to national, ethnic, religious, and linguistic minorities, they have been recognized to maintain their practices and religious activities in an independent way. [5]In 2001 Durban declaration has urged the government of all countries to create favourable conditions for the minority section to express their features within their jurisdiction and to ensure their equal participation in the social, cultural, economic, and political life of the country where they live. These are initiatives by the Government of Canada and the United Nations along with UNCHR. Unfortunately, rules have not been implemented properly in many countries for the protection of minorities. Communalism, fundamentalism, and nationalism are the three most primary causes of the deprivation of minorities in every country. COVID-19 is a new cause through which once again minorities are in great danger from the Coronavirus and from society.

Conclusion: the feature question is what should be done for them? Minorities, refugees, asylum seekers, undocumented workers are facing trouble everywhere in the world. Coronavirus has brought more difficulties for them. Political will force is not really seeking the solution of these community problems through structural equality. They are now more vulnerable due to several restrictions such as moving restrictions implementation by the global leaders. The leaders do not really want migrant people and their global movement for work and other activities. Nationalism rises everywhere, therefore the importance of the citizens has been prioritized rather than the minority community and their other sections. Therefore, where should they go? Who will help them? UNICEF works worldwide to protect the rights of minorities and migrant people. UNICEF is also working for displaced children by providing life protection equipment in refugee camps, child-friendly spaces for safe play. UNICEF and its other nongovernmental organization (NGOs) are working with the government, the private sector, and the civil societies of every country for the solution of migration, minority problems. The Global Refugee Compact is an international agreement that has set up several building blocks for a stronger and positive global response on refugee issues. In 2018 GRC provides a roadmap to various humanitarian international organizations, and host countries for the betterment of minority and migrant communities. They urge them for the integration of the migration community into the mainstream of society. [6]GRC implements four objectives for the development of minorities and migrants. GRC keeps pressures on host countries, increases refugee self-reliance, expansion of resettlement and its access by those community and a safe return of all refugees, migrant temporary workers, or any minority class to their original countries. UNICEF and UNCHR are committing Global Compact on Refugees and make a blueprint for joint action with UNCHR for the reaching goals of migrant's problems solution. The blueprint of the UNICEF with the collaboration of UNCHR is a good document for a better practice around the world to support the children and youth of the refugees, minorities, and migrants. The Global Compact for the safe and regular pathways of migration is a milestone that has recognized the first time that children are central to migration management. UNICEF has worked a lot for the finalization of the document to protect the rights of the children of deprived communities in every country around the world and the children of the migrant community. It is now a big effort to comply with governments for the commitment and bring a real change and positive impact in the lives of children and all minority, migrants. The pandemic makes minorities more vulnerable to others. Therefore [7]UNICEF calls for global action to keep children healthy and nourished and asks the government and partners for sustained life-saving maternal, newborn, and child health services. It has been said that the response of

COVID-19 will have to be strong in health care and ensures universal quality care for all. The organization has given the importance of clean water, sanitation and hygiene for the children and the communities of the minority. It has campaigned about proper handwashing and hygiene practices. The organization has campaigned for child learning. It said that we must do more for the susceptible communities and their children based on equal access to quality learning. The vocabulary Gap is one of the main problems in U.S. society, where children from minorities and migrants are way behind the white children of the United States. The gap is due to structural inequalities between major and minor societies. It is also a result of the impact of vulnerability. Migrants and the parents of the minority have lost their encouragement to build their children strong due to their economic problems. UNICEF and the U.S Board of Education have taken initiatives for equal access to learning modules for all. UNICEF makes sure that social protection such as cash transfer, support for food and nutrition are available for the vulnerable groups, including government assistant and protection of the jobs. There are more initiatives that have been taken by UNICEF. It is working with government and global health partners to making sure that vital supplies of medicine and health equipment are available for the most vulnerable communities, it has(UNICEF) given prioritized delivering lifesaving medicines, nutrition and vaccine, it has given priorities to provide clean water, sanitation and hygiene to the most vulnerable communities, it has provided peer to peer learning and information sharing among the groups like children, adolescence and young people for their mental health support, xenophobia and discrimination.

References

1] *Migrant and displaced children. (n.d.). Retrieved from https://www.unicef. org/migrant-refugee-internally-displaced-children*

2] *Protecting the most vulnerable children from the impact of coronavirus: An agenda for action. (n.d.). Retrieved from https://www.unicef.org/coronavirus/ agenda-for-action*

3] *More Than Meets the Eye - Let's Fight Racism. (n.d.). Retrieved from https://www.un.org/en/letsfightracism/minorities.shtml*

4] *Greenaway, C., Hargreaves, S., Barkati, S., Coyle, C. M., Gobbi, F., Veizis, A., & Douglas, P. (2020, July 24). COVID-19: Exposing and addressing health disparities among ethnic minorities and migrants. Retrieved from https:// academic.oup.com/jtm/advance-article/doi/10.1093/jtm/taaa113/5875716*

Endnote

[1]COVID-19: Exposing and addressing health disparities among ethnic minorities and migrants. Paragraph1

[2]COVID-19: Exposing and addressing health disparities among ethnic minorities and migrants. Paragraph1

[3]More Than Meets the Eye - Let us Fight Racism. Paragraph2

[4]More Than Meets the Eye - Let us Fight Racism. Paragraph 1

[5]More Than Meets the Eye - Let us Fight Racism. Paragraph 4

[6]Migrant and displaced children. Paragraph. Global Refugee Compact. Paragraph 1

[7]Protecting the most vulnerable children from the impact of coronavirus: An agenda for action

COVID-19-Pandemics and Migration in the Context of Canadian Scenario

Ananda Majumdar

Ananda Majumdar

The University of Alberta (Bachelor of Education after Degree Elementary, Faculty of Education*
Community Service-Learning Certificate and Certificate in International Learning, CIL) *

Harvard Graduate School of Education (Professional Education as a Child Development Educator, Certificate in Early Education Leadership (CEEL-Series 2), online) *

Prospective Summer School in History and Archeology (2021-22)-"On the footsteps of Jesus: Jerusalem to Magdala, European University of Rome, Italy

Certificate in Migration Studies, GRFDT, New Delhi, India (Online)

Grant MacEwan University (Diploma in HR Management)

Jadavpur University (Master of Arts in International Relations) Sikkim Manipal University (Master of Business Administration in HR and Marketing Management)

MBB College, Tripura University (Bachelor of Arts in Political Science) Antarctic Institute of Canada (Researcher and Writer), Servicing Community Internship Program (SCiP) Funded by the Government of Alberta Member of Student Panel, Cambridge University Press, Member of the Association of Political Theory (ATP) University of Massachusetts

Student Member of ESA (European Studies Association), Columbia University, U.S.

General Coordinator, Let's Talk Science, University of Alberta

Early Childhood Educator, Brander Garden After School Parents Association

Book Pecker Fellow, Peace X, India

Intern, Migration, GRFDT, New Delhi, India

Cell# 1-780-660-7686

Edmonton, Alberta, Canada

anandamajumdar2@gmail.com, ananda@ualberta.ca

The impact of the Novel Coronavirus (COVID-19) pandemic in the areas of migration, immigrant, populations, gender (women and children) and Canada's immigration and settlement systems have been affected. The global crisis of pandemic differentiates between national citizens and non-citizens, residents, and non-residents etc. It increases social discrimination and inequalities in every country where immigrants, women, children, visible minorities etc. are facing problems and inequalities from the dominated communities. These impacts have happened in the sector of health, economic and social compasses. The closing of international borders, the travel ban has presented various challenges for innocent temporary workers, asylum seekers, refugees, underground migrants etc. As a result, social, economic, and demographic vulnerabilities have been created among the minority groups of every country. Various migrant workers are working in the essential sectors who have a higher risk than other categories of employees in the similar sector because of their poor economic and demographic conditions. They are more vulnerable in the context of losing employment at any time, which created psychological stress among them through losing their houses, less financial capabilities. [1]In this entire scenario, the concept of social resilience is important to understand the relations between pandemics and the vulnerable groups of every country. Resilience can help to understand the impact of the crisis on the societies and communities of those vulnerable classes and the policies of decision-makers. Resilience is also a term that can help to realize the crisis over a system and people and their ability to overcome the crisis through proper management. After a declaration by the WHO on [2]March 11, 2020, COVID-19 has been recognized as a global pandemic. 180 countries have decided to close their borders and imposed a travel ban for global travellers. As a result, the health crisis has been transformed into a mobility crisis around the world because of the travel ban. [3]OECD countries like Canada had taken similar decisions by closing their border with the United States and internationally, along with voluntary travel ban for foreigners. [4]The beginning of the travel restriction was announced by the Canadian Government on March 16, 2020, while the diplomats, family members and United States citizens were exempt from the travel restriction. The restriction was inflicted on asylum seekers who came from the United States and urges for asylum in Canada through every kind of port of entry. The federal government of Canada before decided to provide a chance for asylum seekers for two weeks of quarantine in Canada after their entry from the United States. It was praised by the UNCHR in the context that those seekers can still stay and apply for asylum application in Canada, therefore it was an opportunity yet even a terrible crisis all over. However, later the government reversed their policies and restricted asylum seekers by saying a [5]"temporary measure" in the context of [6]"exceptional times." Prime Minister Justin Trudeau has announced that the border restrictions are [7]"in line with Canada's values on the treatment of refugees and vulnerable people and emerged out of negotiations with the United States to find a mutually acceptable process to deal with irregular migration." In the United States asylum was ended with the Trump Government in the context of their COVID-10 initiative for the protection of Americans. The Centre for Disease Control and Prevention in the United States has announced an order for the deportation of non-citizens who were living in the U.S. temporarily, either legally or illegally. On the other side, the U.S. government took initiatives for the protection of all health issues due to pandemic and the border officials had the authority to [8]bypass immigration law. Estimated 10,000 asylum seekers had been deported across the US-Mexico border throughout the port of entry and their entry had been thus expelled. [9]The deportation was criticized by the UN against the violation of internal law. The collective Caring for Social Justice has criticized the federal government of Canada for their revision of deportation policy and disallowed asylum seekers

18

from the U.S. to Canada in this circumstance. According to them, it was against global harmony and the creation of racism or nationalism by the Government of Canada. Migrants have been characterizing as outsiders and the [10]the transmitter of disease and danger. The border has been transformed into a tool for the creation of orders among various kinds of groups within the Canadian immigration system. Asylum seekers have been recognized as low orders in the hierarchies in both U.S. and Canadian borders. Travel restriction also affected refugees and the state of resettlement in obtaining countries like Canada. [11]On March 17, 2020, The United Nations Refugee Agency (UNCHR) and the International Organization for Migration (IOM) voluntarily announced that the resettlement process of refugees will be suspended due to the pandemic scenario. Resettled refugees who have been already granted relocation from one country to another have been permitted to enter and resettlement process in the hosting countries. Canada is a leading country in the context of the resettlement process since 2018 by surpassing the United States and the United Kingdom through the local number of refugee acceptance. [12]Canadian Government targeted to resettle an estimated 47,950 refugees in 2020 along with estimated 31,950 refugee resettlement in 2021. Refugee camps are always vulnerable due to improper living conditions. Health and medical care and hygiene facilities are limited. Therefore, maintaining social distancing in the refugee camps is exceedingly difficult. Camps in Syria, Bangladesh and Greece are the most densely populated and therefore immediate evaluation of camps in Bangladesh and Greece have been made for access to health services. [13]Several rights advocates and NDP (National Democratic Party) of Canada launches a program that approved many asylum seekers, migrant frontline temporary workers as a permanent residents of Canada. It is an honour for their contribution to Canada in this crisis. The Premier of Quebec has already declared to grant permanent residency of Canada for those front-line workers who are temporary residents in Quebec. The application will be reviewed case-by-case. For the approval of asylum seekers in Canada who works for long-term care, immigration Canada has pointed out to revisit their immigration rules for the grant of permanent residency of these people.

References

Shields, J., & Alrob, A. Z. (2020). COVID-19, *Migration and the Canadian Immigration System: Dimensions, Impact and Resilience*. York: York University. Retrieved from https://bmrc-irmu.info.yorku.ca/files/2020/07/COVID-19-and-Migration-Paper-Final-Edit-JS-July-24-1.pdf?x82641

Endnote

[1] COVID-19, Migration, and the Canadian Immigration System: Dimensions, Impact and Resilience. social Resilience as a Conceptual Frame. Paragraph 1st.

[2] COVID-19, Migration, and the Canadian Immigration System: Dimensions, Impact and Resilience. Canadian Border Closings and the Shutting Out of Asylum Seekers. Paragraph 1st.

[3] COVID-19, Migration, and the Canadian Immigration System: Dimensions, Impact and Resilience. Canadian Border Closings and the Shutting Out of Asylum Seekers. Paragraph 1st.

[4] COVID-19, Migration, and the Canadian Immigration System: Dimensions, Impact and Resilience. Canadian Border Closings and the Shutting Out of Asylum Seekers. Paragraph 2nd.

[5] COVID-19, Migration, and the Canadian Immigration System: Dimensions, Impact and Resilience. Canadian Border Closings and the Shutting Out of Asylum Seekers. Paragraph 3rd.

[6] COVID-19, Migration, and the Canadian Immigration System: Dimensions, Impact and Resilience. Canadian Border Closings and the Shutting Out of Asylum Seekers. Paragraph 3rd.

[7] COVID-19, Migration, and the Canadian Immigration System: Dimensions, Impact and Resilience. Canadian Border Closings and the Shutting Out of Asylum Seekers. Paragraph 3rd.

[8] COVID-19, Migration, and the Canadian Immigration System: Dimensions, Impact and Resilience. Canadian Border Closings and

the Shutting Out of Asylum Seekers. Paragraph 4th.

[9]COVID-19, Migration, and the Canadian Immigration System: Dimensions, Impact and Resilience. Canadian Border Closings and the Shutting Out of Asylum Seekers. Paragraph 4th.

[10]COVID-19, Migration, and the Canadian Immigration System: Dimensions, Impact and Resilience. Canadian Border Closings and the Shutting Out of Asylum Seekers. Paragraph 5th.

[11]COVID-19, Migration, and the Canadian Immigration System: Dimensions, Impact and Resilience. Refugees: Addressing Human Rights and Recognizing the Contributions of Claimants in Canada. Paragraph 1st.

[12]COVID-19, Migration, and the Canadian Immigration System: Dimensions, Impact and Resilience. Refugees: Addressing Human Rights and Recognizing the Contributions of Claimants in Canada. Paragraph 1st.

[13]COVID-19, Migration, and the Canadian Immigration System: Dimensions, Impact and Resilience. Refugees: Addressing Human Rights and Recognizing the Contributions of Claimants in Canada. Paragraph 5th.

Bio: My name is Ananda Majumdar, from the University of Alberta, Harvard Graduate School of Education. I am a researcher, Early Childhood Educator, and Student of Education Elementary and Social Science. My research background is in Social Science and fifty papers have been published. My research is under the project of the Government of Canada, called, Riipen. I am working at the Antarctic Institute of Canada as a writer and researcher. Degrees like BA, B.Ed., MA, MBA are my best academic achievement. My certifications and references in social science have reached in total of 230, which provides me hands-on experience in academia. As an Education and Social Science student, I pursue a keen interest in exploring the different concepts and facets of these two subjects and their intersecting areas. I am passionate about writing and researching various topics related to Education and Social Science phenomena.

Fetal Alcohol Spectrum Disorder: Biological and Psychosocial Factors

Aleefa Devji, Jasrita Singh, Austin Mardon

Fetal Alcohol Spectrum Disorder, commonly known as FASD, refers to the broad spectrum of disabilities and negative consequences that stem from exposure to alcohol in utero. Although FASD affects roughly 2-5% of the population, it is largely an invisible disorder that makes it hard to recognize (Brown et al.). The physiological, behavioural and neuro-cognitive disabilities associated with FASD make this disorder often difficult to understand as the brain damage that cannot be seen (Bartlett & Lerose, 2005).

The wide spectrum of disabilities that are visible in individuals with FASD can be attributed to factors such as the quantity of alcohol consumption by the mother during pregnancy and the associated time frame (Yuzwenko, 2009). As a result of the varying degrees of disability and factors which may inhibit development of the child, FASD can present itself in the form of distinct facial features or delayed cognitive functions such as problem-solving deficits, difficulties with abstract thinking, or the inability to connect actions with consequences (Dennett, Jonsson & Littlejohn, 2009).

Biological Effects of Alcohol on Prenatal Development

FASD encompasses a spectrum of disabilities which stem from organic brain damage (Coriale et al., 2013). In each individual, different areas of the brain will be affected by alcohol exposure, and as a result, the impact on behaviour and function will be different for everyone experiencing FASD (Badry & Bradshaw, 2010). In general, those experiencing the disorder will experience issues with planning and impulsivity, inherently be very concrete thinkers and more often than not, have trouble connecting concepts and actions (Yuzwenko, 2009).

It is helpful to understand the way that alcohol affects a developing brain, and the way that a normal brain will develop in the absence of these teratogenic factors, agents present in the surrounding environment that can disturb the development of an embryo or fetus (Coriale et al., 2013). When in utero, the normal brain is developing as it grows outward from the core. However, when alcohol is introduced during the process of outward structural development, the nerve cells often become confused and will make connections to other nerve endings in the wrong places (Dennett, Jonsson & Littlejohn, 2009). As a result, the brain will become disorganized at a basic level. The brain damage associated with FASD may also be attributed to failure of the brain developing, and the tendency to see cell tissues dying when exposed to alcohol during development (Coriale et al., 2013).

An easy way to imagine this disorganization of connections is the analogy of a gumball machine representing a brain. Each colour gumball represents a category of information, although the gumballs are not all grouped together and therefore it is difficult to make connections between them. The lack of organization and the inability to form connections between the coloured gumballs is a very simplistic representation of the nerve cells in the brain that are unable to connect, and the areas of information that are not organized as they should be.

Functions of the brain that are related to the inward core of the brain can be attributed to the Limbic brain and the Neocortex respectively (Dennett, Jonsson & Littlejohn, 2009). Having a better sense of the functions of each area will enable a better understanding of the difficulties that individuals with FASD are facing. Tied to the Limbic brain is the emotional learning, anger and sexual emotion, the functions of which are often preserved, although this area of the brain is commonly thought of as a gas pedal to the brain (Badry & Bradshaw, 2010). For this reason, many FASD individuals will have issues controlling sexual impulses, anger and emotions when dealing with difficult situations (Pei, 2010). On the outside layers of the brain, the Neocortex is tied to higher-order thinking, complex thought, problem solving and judgment or reasoning. Alcohol's effect on the disorganized nerve connections made in the outer layers of the brain leads to many issues that individuals with FASD face such as trouble processing thoughts, or understanding complex situations (Dennett, Jonsson & Littlejohn, 2009).

Specific areas of brain development may also have negative consequences on the actions and abilities of those with FASD. In general, the damage to the brain may result in individuals being incapable of understanding societal rules and expectations (Brown et al.). For instance, the damage to the corpus callosum would affect the ability of individuals to put rules and instructions into play (Dennett, Jonsson & Littlejohn, 2009). Damage to the cerebellum and basal ganglia may affect cognitive processes like attention or result in issues with perception of time. Damage to the frontal lobe of the brain significantly affects executive functioning (Bartlett & Lerose, 2005). This includes difficulty connecting cause and effect, trouble learning from their mistakes and the inability to generalize learning in different situations.

The effects of each of these impairments can be detrimental to social functioning on their own, many FASD individuals deal with these challenges at the same time. The compilation of cognitive disorders experienced by many FASD affected individuals often lead to problematic behaviour in society, which result in trouble with the school system and perpetually with authorities. Therefore, it is important to truly acknowledge and understand the developmental issues and cognitive function of FASD affected individuals in order to provide them with a means of equitable treatment.

References

Badry, D. & Bradshaw, C. (2010). Assessment and Diagnosis of FASD Among Adults. A *National and International Systemic Review*.

Coriale, G., et al. (2013). *Fetal Alcohol Spectrum Disorder (FASD): neurobehavioral profile, indications for diagnosis and treatment.*

Dennett, L., Jonsson, E., & Littlejohn, G. (2009). *Fetal Alcohol Spectrum Disorder (FASD): Across the Lifespan.*

Knowledge Network (Producer) Bartlett, S. & Lerose, M. (Director). (2005). F.A.S. *When The Children Grow Up* [Motion Picture]. Canada.

Pei, J. (Presenter). (2010). *FASD and Practice: Issues for Probation Officers* [Motion Picture]. (Available from the Government of Alberta).

Yuzwenko, P. (Presenter). (2009). *FASD and the Criminal Justice System: Issues for Defense* [Motion Picture].

Mental Health and COVID-19 in Canada: What is the Cost of Lockdowns?

Jessica C. Henschel

In the past year, everyone across the globe has felt the weight of the novel coronavirus pandemic, COVID-19. Drastic public health measures were taken to combat the spread of the virus and reduce the strain on hospitals. Many countries opted for severe lockdowns and restrictions that severely limited social interaction to members of a household and only allowed citizens to leave their homes for groceries and medical appointments. Despite the beneficial effect these lockdowns had for stopping the spread of the virus, it came at the cost of mental and social health. This left us all wondering: are lockdowns worth the cost of negative mental health outcomes of Canadian residents? This article will explore the variety of public health orders employed in Canada since March 2020 and how these restrictions have impacted Canadians' psychological well-being.

When the first case of COVID-19 was recorded in Canada on January 25, 2020, public health officials across the country began to prepare for the worst (Bronca, 2020). Patient zero — a man who had recently returned from the outbreak epicenter, Wuhan, China to Toronto — was immediately put into quarantine and major airports began introducing extensive screening measures. In February 2020, China's numbers were rapidly increasing, and Canada began seeing a rise in travel-related cases. The exponential growth of cases in March were enough to see almost every province declare a state of emergency (Bronca, 2020). All retail, gyms, recreation centers, restaurants, casinos/banquet halls, and concert venues were shut down. In the height of the first wave (March-June 2020), citizens were only allowed to go out for groceries and medical visits. Social gathering were prohibited, and physical contact was limited to household members only (Karstens-Smith, 2020). Due to the public health acts, it was illegal to break the lockdown rules and would result in fines or jail time. This came as an understandable shock and brought about an innate sense of loss as people across the world were laid off, forced to close their businesses, could not attend school, and most significantly, not see their friends or family. As Karstens-Smith (2020) reported for Global News, the new restrictions were met with outrage and tremendous fear. Unfortunately, none of us predicted that we would still be living in these conditions over one year later.

Due to the loss of lives and livelihoods during the COVID-19 pandemic, the mental health of Canadians has been negatively impacted. Mental Health Research Canada (MHRC) released that there had been a significant increase of adult Canadians reporting depression (22%) and anxiety (20%) disorder diagnoses (Flanagan, 2021). These numbers were the highest rates MHRC had ever measured, with 6% of Canadians (1.8 million) having all 4 negative indicators that polling uses to track mental health: high anxiety and depression, moderate to severe mental health symptoms, low management of stress, and low resiliency (MHRC, 2021). In addition to MHRC's survey, the Canadian Mental Health Association (CMHA) partnered with University of British Columbia to provide a detailed study of the effects of COVID-19 on psychological health in Canada (The University of British Columbia [UBC], 2020). They reported that the second wave of the pandemic brought about increased levels of stress and anxiety. About 71% of Canadians stated they were worried about the continued waves of the pandemic, with 58% being worried about the mortality of their loved ones (UBC, 2020).

Another concerning factor is that research has revealed a negative impact on family and child mental health. Gadermann et al.'s (2020) study on the impacts of COVID-19 on Canadian mental health found that families with children under 18 have experienced a deteriorated mental health due to the pandemic. Significantly, parents reported that their children's mental health had tremendously worsened since the start of the pandemic and their interactions were frequently more negative than positive — operationalized as greater conflicts occuring within the family unit. Additionally, more parents compared to the rest of the sample reported increased alcohol and substance consumption, suicidal thoughts/feelings, and worry about being safe from physical/emotional domestic violence (Gadermann et al., 2020).

Of greatest concern to mental health service providers and the general population was the sharp increase in suicidality since the COVID-19 pandemic began. With 1 in 10 Canadians experiencing feelings or thoughts of suicide, public health researchers noticed the severe psychological effects of the virus and restrictions (UBC, 2020). Suicidal ideation was even stronger amongst already marginalized groups, such as those who identify at LGBTQIA2S+, individuals with pre-existing mental illness, disabilities, and Indigenous peoples. Mental health and addictions researcher at UBC, Emily Jenkins stated, "we are seeing a direct relationship between social stressors and declining mental health. Particularly concerning are the levels of suicidal thinking and self-harm, which have increased exponentially since before the pandemic [...]" (UBC, 2020). One of the primary stressors Jenkins referred to is the financial strain from the pandemic. More than a third of Canadians were concerned about finances, with the majority of individuals being parents with small children. Many parents are burdened with providing for their families and being able to put food on the table while they have lost their jobs and stability due to lockdowns.

A study by McIntyre and Lee (2020) used macroeconomic indicators (primarily unemployment) from the COVID-19 pandemic to predict suicide statistics in Canada. Based on the suicide and unemployment rates from previous years, McIntyre and Lee (2020) calculated various mortality projections for 2021. If unemployment continues to rise in the country at the same rate as 2020, the projected suicide rate would increase to 13.6% in 2021. This would result in an additional 2,114 suicides across the country. Devastatingly, these deaths could be preventable if Canada had handled the pandemic response in a different manner in March 2020. Therefore, these numbers merit the question: are lockdowns worth it? The economic costs of lockdowns seem to far outweigh the health benefits in some cross-country studies (Thom, 2021). Seemingly the loss of life from suicide may be preventable as well. Despite the mental health crisis in the world, lockdowns may be the only feasible option in preventing strain on healthcare systems and hospitals. Further research and policy work must be done to ensure that the physical *and* mental health of citizens are being protected, while still doing what is in the best interest of the economy.

Due to the current mental health crisis in Canada, there are several supports that have been made available. The Mental Health

Commission of Canada has created a program called Wellness Together Canada, which provides 24/7 mental health support to Canadians (Mental Health Commission of Canada, 2021). Wellness Together Canada provides various services: access to free, live counselling through either text (SMS) or phone, addiction help, search for psychological resources in your area, and self-care information/activities. Every province and territory has set up their own government supports as well, such as Here to Help British Columbia, Help in Tough Times Alberta, Care for Your Mental Health Manitoba, COVID-19 Support for People Ontario (Mental Health Commission of Canada, 2021). If you are experiencing suicidal ideation or struggling with your psychological well-being, please contact your nearest government support or call 911 for a mental health emergency.

The COVID-19 pandemic has affected us all in a variety of ways and mental health is one that has been overlooked. Lockdowns are one of the only ways to control the spread of the virus and governments are struggling to find solutions. Thanks to the innovation of major biotechnology companies, vaccines are being produced. There is hope for a better future and everyone has worked hard to limit contagion. Canadians still require support and much needed social connection to stay vigilant and resilient. Take advantage of local organizations who are there to help and continue to keep one another safe.

References

Bronca, T. (2020, April 8). *COVID-19: A Canadian timeline.* Canadian Healthcare Network. https://www.canadianhealthcarenetwork. ca/covid-19-a-canadian-timeline

Flanagan, R. (2021, January 14). *Canadians reporting more anxiety and depression than ever before, poll finds.* CTV News. https:// www.ctvnews.ca/health/coronavirus/canadians- reporting-more-anxiety-and-depression-than-ever-before-poll-finds-1.5266911

Gadermann, A.C., Thomson, K.C., Richardson, C.G., Gagné, M., McAuliffe, C., Hirani, S., Jenkins, E. (2020). Examining the impacts of the COVID-19 pandemic on family mental health in Canada: findings from a national cross-sectional study. *BMJ Open, 2021*, 1-11. https://doi.org/10.1136/bmjopen-2020-042871

Karstens-Smith, B. (2020, December 28). *A timeline of COVID-19 in Alberta*. Global News. https://globalnews.ca/news/7538547/covid-19-alberta-health-timeline/

McIntyre, R.S., & Lee, Y. (2020). Projected increases in suicide in Canada as a consequence of COVID-19. *Psychiatric Research, 290*, 104-113. https://doi.org/10.1016/j.psychres.2020.113104

Mental Health Commission of Canada. (2021). *Government of Canada COVID-19 resources.* https://www.mentalhealthcommission.ca/English/government-canada-covid-19-resource

Mental Health Research Canada [MHRC]. Mental health in crisis: How COVID-19 is impacting Canadians. https://www.mhrc.ca/national-poll-covid/findings-of-poll-5

The University of British Columbia. (2020, December 3). *New national survey finds Canadians' mental health eroding as pandemic continues.* https://www.med.ubc.ca/news/new- national-survey-finds-canadians-mental-health-eroding-as-pandemic-continues/

Thom, H. (2021, February 10). *Lockdown critics are sure the costs outweigh the health benefits, but they're wrong.* Phys.org. https://phys.org/news/2021-02-lockdown-critics-outweigh-health-benefits.html

Difficulties in Diagnosing FASD in the Judicial

Aleefa Devji, Jasrita Singh, Austin Mardon

Despite the fact that we are still in the pandemic, multiple vaccines have been developed and approved, and shipments have been trickling into Canada. However, The lag in Canada's vaccine rollout has been frustrating for many. Even though the early Canadian COVID-19 responses have outshined our southern neighbour, it now seems like the US will emerge from this pandemic much earlier than Canada. As of March 10, Canada ranks 41st in the world for the number of people vaccinated per 100 people, with 6.93 people vaccinated per 100 Canadians. For comparison, the US sits in 7th place with 28.83 per 100 people, and Israel beats out the rest of the world at 100.07 per 100. In fact, recently, the US averaged more vaccinations in one day than the total number of people vaccinated in Canada. While the vaccine rollout in the US edges us out now, it wasn't always like that and they have had their share of bumps along the way. From their initial hurdles, a new phenomenon, vaccine hunting, came into existence, and controversial as it may be, it could very well be the answer Canada is looking for.

The American vaccination campaign did not get off to a great start, to say the least, leading to shortages, confusion, and wasted vaccines. In Florida, for example, the lack of a centralized roll-out program led to long lines and confusion, as each county makes a different roll-out approach. This confusion leads to waste when an expired vaccine cannot be used. It is within that chaos that vaccine hunting emerged. The vaccine hunting described here is not like the well-publicized events of wealthy individuals flying abroad or to remote regions to jump the line and take vaccines away from vulnerable communities. No, the vaccine hunting described here refers to a phenomenon where individuals stake out a vaccination site for long hours and wait to take their left-over vaccines that are set to expire. The act of doing so is controversial, as many vaccine hunters are young, and can afford to spend the time, money, and energy, to spend an entire workday waiting for a vaccine

31

that may or may not be available.

Many are uneasy about vaccine hunters, seeing them as healthy young people jumping the line, taking away vaccines from the most vulnerable. However, the unfortunate truth is that if they weren't doing it, the vaccines may very well end up as medical waste rather than in someone's arm. The Pfizer vaccine, for example, must be stored at -70°C for storage, and prior to use, it needs to be thawed to 2-8°C, at which point it can be stored for 120 hours. The vial then needs to be diluted before injection, and once that occurs, there is only a 6-hour window to use up that vial before it needs to be discarded. During a massive snowstorm in Oregon in late January, healthcare workers were stranded on the road along with six doses of the Pfizer vaccine due to expire. The healthcare workers proceeded to hand out the vaccines car by car until they were all gone. This has the same basic underlying concept as vaccine hunting, and yet it is depicted with much less controversy.

Although vaccine hunters are seen to be in a moral grey zone in North America, Israel has embraced its underlying concept and used it to run arguably the most successful mass vaccination campaign during this pandemic. Israel vaccinated 90% of their population in less than three months, and one of the many things that it did right was flexibility with vaccine distribution guidelines. At their mass vaccination centers, after the daily appointments are over, which consisted mainly of vulnerable groups initially, the workers would go on the streets asking anyone they encounter if they would like a vaccine, they had leftover, regardless of priority status. This ensures that as many people would be vaccinated as possible, using all their stockpiles to the full potential. This practice may not have been the key driver of their success, but it nonetheless played a part.

The flexibility displayed in Israel has also been displayed to a limited degree in the US, as a man from DC received a dose of vaccine from an open vial when he was simply getting groceries. There have even been Facebook groups made to connect vaccination centers with expiring doses and the "vaccine hunters". This allowed people to efficiently bypass the official distribution schedule and organize these hunting trips. This form of "organized" or "centralized" may be a way to make mass vaccination more efficient. It can fill the gaps in the government roll-out system to make sure the least number of vaccines go to waste and the greatest number of lives get saved, regardless of who those lives are.

Now in Canada, with several large shipments of the Pfizer and Moderna vaccines on the way, the tentative loan of 1.5 million doses of the AstraZeneca vaccine from the US, and the likely approval of the Johnson & Johnson vaccines, having a flexible vaccine rollout program will mean more people vaccinated, more lives saved, and a faster return to normalcy. Any rollout program, with "vaccine hunting" or not, needs to also prioritize the number of people vaccinated, not just who is being vaccinated. While no rollout system is perfect, wouldn't it be better if a vaccine ended up in the arm of a 20-year old, rather than not going in an arm at all?

AUTHOR AFFILIATIONS:

Cheng En Xi: Faculty of Health Sciences, McMaster University, 1280 Main St W, Hamilton ON, L8S 4L8, Canada

Cheng En Xi is a member of the Bachelor of Health Sciences (Honours) program at McMaster University.

Austin Mardon: Faculty of Medicine & Dentistry, University of Alberta, 2J2.00 WC Mackenzie Health Sciences Centre, 8440 112 St NW, Edmonton AB, T6G 2R7, Canada.

Austin Mardon is an Assistant Adjunct Professor at the Department of Psychiatry & John Dosseter Health Ethics Centre of the University of Alberta, as well as a Special Advisor, with the Glenrose Rehabilitation Hospital.

Fetal Alcohol Spectrum Disorder, commonly known as FASD, refers to the broad spectrum of disabilities and negative consequences that stem from exposure to alcohol in utero. The spectrum of developmental disabilities that FASD individuals face are cognitive, physical, and psychosocial, while the severity of the delays and developmental changes are a result of the amount of exposure and the time of exposure to alcohol in utero.

As a relatively new concept to the justice system, and specifically legal professionals, there are many unanswered questions in regard to the prevalence, diagnosis and long-term effects of FASD. It is considered a mental disorder that must be diagnosed by a physician, although a diagnosis can only take place if there is admission from the mother that she consumed alcohol during pregnancy (Pei, 2010). As a result, many individuals with FASD go undiagnosed. This has been a difficult issue for physicians, due to the stigma surrounding the overuse and abuse of alcohol, which will likely prevent individuals from coming forward and admitting their alcohol use during pregnancy.

In any case, the diagnosis of FASD is best undertaken by a team of individuals to focus on different aspects of the individual. Two key areas for assessment during diagnosis are facial features, and brain development or injury (Badry & Bradshaw, 2010). The latter gives more information to the prevalence of FASD with the help of brain scans, which enable doctors to make the connections between alcohol affected brain development and the behaviors observed on a daily basis. Typical facial dysmorphologies, on the other hand, are not very useful. The face develops over a short three-day period, from the 9th to 21st day of gestation, which means that not all children or adults will display the commonly observed features (Dennett, Jonsson & Littlejohn, 2009). Those who do have the distinct smaller eyes, flatter noses and thin upper lips may also outgrow them by adulthood. For this reason, the absence or prevalence of these features cannot stand alone to diagnose individuals with FASD. The necessity for a team or physicians and the use of advanced technology makes a diagnosis for FASD costly, and this is the primary reason that many individuals go undiagnosed.

The behaviours exhibited by individuals with FASD are quite common to other behavioural diagnosis such as Oppositional Defiance Disorder (ODD) and Attention Deficit Disorder (ADD). ODD is a diagnosis characterized by irritability, defiance and anger management issues, which are quite common in individuals and particularly youth with FASD. As a result, it is common that, without knowing whether alcohol was consumed during prenatal development, a doctor will attribute the behaviour to ODD or ADD as a behavioural diagnosis (Brown et al.). This is also especially common within Caucasian middle-class families who, due to social expectations, fail to admit to the consumption of alcohol during pregnancy.

FASD is particularly difficult to diagnose with youth in the child welfare system. Researchers have found that 80% of FASD youth were not raised by their biological mothers, making it difficult for doctors to definitively diagnose FASD (Bartlett & Lerose, 2005). Youth with FASD who are also within the child welfare system are 60% more likely to end up entrenched in the criminal justice system, where their chances of being diagnosed are again quite slim (Yuzwenko, 2009). Many of the professionals associated with the criminal justice system are not trained to recognize FASD and as a result these individuals, who are undiagnosed upon entry, fall through the cracks.

Youth with FASD deal with adverse effects of the disorder and due to the poor likelihood of having a definitive diagnosis, they also lack the necessary support to deal with them. As well, it has been found that one of the strongest correlations of these adverse outcomes is a lack of an early diagnosis. Children diagnosed before the age of eight have a better chance at a successful life and establishing effective supports which will largely affect their chances of becoming a successful adult (Dennett, Jonsson & Littlejohn, 2009). When we begin to attempt diagnosing individuals later in life, the skills they are working with are less pronounced and the features that are present will likely have softened or been covered by secondary disabilities (Native Counselling Services of Alberta). As well, it is more likely that individuals are further removed from the possibility of confirming their exposure to alcohol during their prenatal development.

The FASD Affected Individual's Journey Through the Justice System

FASD has a 10 times greater prevalence in the correctional population than in the general population, which means that approximately 20% of individuals who are incarcerated have

FASD. Although this is the case, it is only recognized in some court proceedings, "that FASD is a disability that reduces the moral culpability or voluntariness of a person's actions" (Institute of Health Economics, 2013, p. 4). A study conducted in New Brunswick also discovered that only about 40% of the judges and 26% of prosecutors reported feeling prepared to deal with cases related to individuals with FASD (Cox, Clairmont & Cox, 2008, p. 309). This case is only truly representative of one province, although it can be assumed that as a result of similarities between provincial regulations and education that the findings are representative of the need for FASD education throughout the nation.

FASD affected individuals who enter the criminal justice system are often lost upon entry. From the point of being charged, they may be unable to grasp their rights, or they may refuse a lawyer's presence if they lack the understanding of having done something wrong.

Without having the necessary training]for dealing with FASD individuals, it is not possible for officers to be certain that a youth understands their rights, or what implications their actions might result in. In dealing with initial statements from individuals with FASD, it is also important to understand their desire to please others and their ability to be taken advantage of by others (Bartlett & Lerose, 2005). These characteristics, commonly seen in FASD individuals, can lend them to give false confessions or to give inaccurate accounts for the events that have taken place.

FASD individuals often have difficulties understanding legal jargon when engaging with lawyers or judges. Understanding probation conditions, no-contact orders or comprehending the evidence against them can be difficult, if not impossible. In instances such as these, it is the legal professional's responsibility to ensure understanding of the processes. If not, the individuals are being set up for failure from the start. McLaughlan et al. (2014) argue that:

> "Intervention recommendations designed to optimize learning for offenders with FASD might include simple language, repeating information, ensuring attention is captured before communicating information, [and] gauging comprehension frequently to assess adequacy of learning. (p. 19)"

Criminal and Sexual Activity Associated with FASD

Aleefa Devji, Jasrita Singh, Austin Mardon

Fetal Alcohol Spectrum Disorder, commonly known as FASD, refers to the broad spectrum of disabilities and negative consequences that stem from exposure to alcohol in utero. The spectrum of developmental disabilities that FASD individuals face are cognitive, physical, and psychosocial, while the severity of the delays and developmental changes are a result of the amount of exposure and the time of exposure to alcohol in utero.

Youth with FASD compared to youth without a mental disorder have difficulty with delayed gratification, experience less inhibitions, and cannot always appreciate the consequences of their actions (Mitten, 2004). More often than not, we see FASD individuals engaging in crimes related to property, assault and theft, as well as crimes that are often impulsive, opportunistic and which fail to make sense (Native Counselling Services of Alberta).

When the average individual without a mental health disorder is reprimanded, they are less likely to repeat similar behaviours and are more likely to act with an understanding of their behaviour. Contrastingly, the average fetal alcohol affected individuals are not able to understand the way that consequences have actions as in cause/effect relationships and this encompasses their inability to generalize what they learn from one situation to another (Yuzwenko, 2009).

As a result of their inability to learn from their mistakes or understand the magnitude of their crimes, these FASD affected youth and adults are likely to be repeat offenders. While the criminal justice system is based on a notion of criminals being caught, punished and learning from their mistakes, failure to account for the unique needs of FASD individuals means that they experience the perpetually revolving door of the criminal justice system. The accused is expected to have an

37

intellectual understanding of their wrongdoing, but with the invisibility of FASD as a disorder, and its lack of diagnosis, their behaviour is seen as willful and defiant.

The inability for youth to generally differentiate between right and wrong is disclosed by findings that approximately 50% of all subjects with FASD within a study, who are older than 12 years of age, have engaged in inappropriate sexual behaviour (Chartrand & Forbes-Chilibeck, 2003). In regard to sexual impulsivity and inability to delay gratification, sexual encounters are often difficult for individuals with FASD to understand as the average person would. Judge Stuart has suggested that the propensity for sexual misbehaviour in individuals with FASD may be a result of physiological aspects of the disorder which result from lesions in the left frontal lobe, caused by brain damage associated with the alcohol exposure (Chartrand & Forbes-Chilibeck, 2003). He has also proposed that the condition of FASD creates this generalized lack of ability to exercise impulse control or to inhibit the immediate gratification of impulses. In Fraser J's theory of how FASD, sexual urges and impulsive behaviour interact, (cited in Chartrand & Forbes-Chilibeck, 2003) he states:

> "Herein lies the problem relating to the commission of sexual offences. Having a mature body beyond its intellect, he has urges for sexual gratification which leads to impulsiveness and unpremeditated behaviour without caution and with risk taking. This is followed by non-comprehension that the behaviour was inappropriate."

This ideal of thought reveals that individuals with FASD might end up victimizing others without fully understanding the magnitude of their actions. The diminished understanding of what is appropriate and what is not may prevent these individuals from seeing the importance of mandatory consent involved with sexual acts and will often lead to criminalization for their actions. This concept of an act being criminal without consent, but legal with consent, can be confusing for young people dealing with a mental disorder, who are also experiencing many psychological and physiological changes associated with puberty.

In terms of females with FASD and sexual activity, a study by Streissguth et al. (2004) found that 94% of the females who have engaged in inappropriate sexual behaviours have also experienced sexual assault, physical abuse or violence. Of these women, 57% were also found to have alcohol or drug abuse problems. The high rates of sexual offences linked to females affected by FASD, who also engage in substance abuse, increases the likelihood that these women will give birth to children affected by FASD. Research by Totten (2010), has also discovered that mothers of FASD affected children are often victims of physical and sexual abuse as children, experience violence during their pregnancies, and suffer serious mental health problems. As a result, social systems need to address underlying factors of the addictions issues in order to lower the prevalence rates of children born with FASD or to educate caregivers, healthcare workers, and the criminal justice system about the unique challenges posed to FASD affected individuals.

References

Chartrand, L. N. & Forbes-Chilibeck, E.M. (2003). The sentencing of offenders with Fetal Alcohol Syndrome. *Health Law Journal*, 11, 35-70.

Mitten, R. (2004, January). "Fetal Alcohol Spectrum Disorders and the Justice System" (Section 9). From The First Nations and Métis Justice Reform Commission Final Report, Volume II.

Native Counselling Services of Alberta (Producer). (2010). *Invisible: Fetal Alcohol Spectrum Disorder and the Justice System* [Motion Picture].

Streissguth, A. P., Bookstein, F. L., Barr, H. M., Sampson, P. D., O'Malley, K., & Young, J. K. (2004). Risk factors for adverse life outcomes in fetal alcohol syndrome and fetal alcohol effects. *Journal of Developmental and Behavioral Pediatrics*, 25(4), 228-238.

Totten, M. (2010). Investigating the linkages between FASD, gangs, sexual exploitation and women abuse in the Canadian aboriginal population: A preliminary study. *First Peoples Child & Family Review.* 5(2), 9-22.

Yuzwenko, P. (Presenter). (2009). *FASD and the Criminal Justice System: Issues for Defense* [Motion Picture].

The Discoveries That Led to Bacterial Specificity and Bacteriology

Aleefa Devji, Jasrita Singh, Austin Mardon

Before the study of Bacteriology began in 18th century Europe, it was not fully understood how contagion and infection were spread through populations. During the Bubonic Plague (The Black Death) in the 14th century, it was a common conception that epidemics arose as a result of punishment for sins and other natural calamities[1]. It was not until the discovery of bacteria by Anthony van Leeuwenhoek in the 17th century that there was a drastic shift in medical, and microbiological understandings. Further experimentation on bacteria by Louis Pasteur—whose curiosities grew from previous experimentation with fermentation and putrefaction— led to the understanding of distinctive microorganisms causing infection and spread of disease. The subsequent study of bacteriology which arose from this experimentation and fundamental research which paved the way for development of vaccines and allowed for disease prevention and treatment as well as a greater scientific understanding of epidemics such as the plague.

In the early 1300s, the pandemic that was the Black Death — now known as the Bubonic Plague — swept across Europe and Asia, killing millions[2]. In the 14th century, the Black Death was believed to be a cause of religion or spirituality casting punishment on those who had sinned. Alternatively, it was thought to be a result of natural disasters such as floods, droughts, or famine. In some cases, it was also commonly perceived that illness was a cause of imbalance within the four humours: blood, yellow bile, black bile, and phlegm. Hippocrates posited that pain or illness was a result of excess or deficiency in one of the four humours[3]. Humoralism gained popularity with the circulation of the Hippocratic Corpus and remained a theory of medicine for centuries after the influence of Galen's writings (129-201 AD). Galen believed that humours were formed in the body and that the ingestions of different foods, geographical landscapes, and environmental differences were able to influence the production of certain humours[4]. In the early years

40

of the Plague, the idea of herbal medicines was adopted by the British to treat the fevers and other ailments although no one was able to suggest a cure for the plague. Coined by some as the "new disease"[5], the plague was untreatable because it was not fully understood and for this reason, it was thought that the best way to manage the outbreak was to leave the contaminated areas. It was not until much later in scientific discovery that bacteria and microorganisms were known and understood to better treat illness and disease.

The plague is one of history's most deadly diseases. In the 1300s it was unknown the cause of this disease or how to prevent and treat the symptoms and ailments that were caused as a result. Bacteriology has since created a foundation for discovery and identification of the species of bacteria responsible for the outbreak. At the time of the plague, when bacteriology and spread of disease was not yet understood, there was no way to prevent the mass outbreaks or to properly treat the infected individuals. Symptoms of the plague were not clear and as a result, all cases of illness and fever were treated as the plague. Today, as a result of bacteriology we are able to identify disease and treat each case for its own set of symptoms and we are better able to understand the necessary precautions to prevent spread of disease, bacteria and minimize contagion.

"Animalcules" as Anthony van Leeuwenhoek originally termed bacteria, were first seen the in the 17th century with the invention of the microscope[6]. Leeuwenhoek's first discovery of the appearance of animalcules was in rainwater, snow water, or well water that had organic materials such as cloves, nutmeg or ginger added. Leeuwenhoek also noted that animalcules or bacteria were not present in rain or fresh water when collected, but began to appear over time, particularly with the added organic materials. Although he saw the emergence of these microorganisms, he did not continue further experimentation and understanding of their chemical nature. Leeuwenhoek was instead responsible for the first clear description of animalcules and their morphology and he also found similar protozoa in the scrapings from his teeth, demonstrating that they were present in various environments[7]. Before Leeuwenhoek's discovery of animalcules, there were discussions of contagion and infection, but these microorganisms were not yet seen as causative agents of disease. New experimental methods were required to observe their nature and chemical effects[8]. As a result, the necessity for fundamental research in biological and life sciences has proven to be instrumental in understanding bacteria and the interactions of micro-organisms with higher organisms and humans. The ability for bacteriology to describe and identify these micro-organisms through the process of experimentation and observation furthered the medical field and ability for doctors to identify and treat disease.

Leeuwenhoek's optimization of (clever new use of) the microscope and discovery of microorganisms eventually led to an increase in support of disease caused by parasites and bacteria in the 17th century[9]. The microscope allowed scientists to further observe and understand of a new class of organisms, invisible to the naked eye. This discovery eventually led to the field of medical bacteriology; the study of causative agents of infectious disease and their effects on the human body as well as the learnings of diagnosis, treatment, and prevention associated with the disease[10].

Scabies was the first infectious disease that was physically seen by humans. Scabies was discovered to be the invasion of a microparasite that burrowed under the skin by a tiny mite, barely visible to the naked eye. The use of the microscope advanced the learnings of Giovanni Bonomo who observed the mite in 1687 to understand its nature and ability to transmit infection which led to better prevention and treatment[11]. The wider significance of Bonomo's work was not greatly understood in the 17th century, although his work was one of the first examples of how the knowledge of etiology can guide justified means of treatment and prevention of disease.

With infectious diseases now visible and better understood to be a result of interactions of micro-organisms and bacteria, there was support for the idea of bacterial and disease specificity. Leeuwenhoek had contributed morphological descriptions of micro-organisms, and Bonomo was able to provide a visible case of the interaction of a unique micro-organism interacting with a human to cause disease. The observations in the 17th century eventually led to Pierre Bretonneau founding the Doctrine of Specificity of Disease in the 19th century. His interest in specificity was piqued by characteristic blister types which were produced by secretions of different beetle species. His concept of specificity came from his study of two distinctive species of bacterium. The first was Diptheria, a species that no matter the type always resulted in symptoms of croup that were different from other throat conditions. The second was Typhoid fever, which was characteristic of lesions in the small intestine[12]. Bretonneau's work with these two distinctive bacteria and the characteristic symptoms they caused, supported the idea that distinctive bacteria affect higher organisms in specific ways which demonstrated specificity in the types of bacteria and the infections or diseases that they caused. His doctrine and beliefs thus addressed the idea that each unique class of bacterium or "morbid seed" caused a specific disease as a result of specificity[13].

These discoveries eventually led to the understanding that bacteria are specific to the diseases they cause and allowed doctors and scientists to cure disease and treat illness. Paving the way for symptomatic presentation to lead the course of treatment chosen, and eventually for the discovery and invention of vaccines to treat and prevent the spread of specific illnesses.

Bibliography

Echenberg, Myron J. *Plague Ports: The Global Urban Impact of Bubonic Plague,1894- 1901*. New York: New York University Press, 2007.

Foster, W. D. *A History of Medical Bacteriology and Immunology*. London: Heinemann Medical, 1970.

Hippocrates., John Chadwick, G. E. R Lloyd, and W. N Mann. *Hippocratic Writings*. Harmondsworth: Penguin, 1983.

Leadbetter, Edward R., and Jeanne S Poindexter. *Bacteria In Nature*. New York: Plenum Press, 1985.

Lindberg, David C. *The Beginnings of Western Science: The European Scientific Tradition In Philosophical, Religious, and Institutional Context, Prehistory to A.D. 1450*. 2nd ed. Chicago: University of Chicago Press, 2007.

Endnote

[1]Echenberg, Myron J. *Plague Ports: The Global Urban Impact of Bubonic Plague,1894- 1901*. New York: New York University Press, 2007.

[2]Echenberg, *Plague Ports*, 21.

[3]Hippocrates., John Chadwick, G. E. R Lloyd, and W. N Mann. *Hippocratic Writings*. Harmondsworth: Penguin, 1983.

[4]Lindberg, David C. *The Beginnings of Western Science: The European Scientific Tradition In Philosophical, Religious, and Institutional Context, Prehistory to A.D. 1450*. 2nd ed. Chicago: University of Chicago Press, 2007. 5 Echenberg, Plague Ports, 53.

[6]Leadbetter, Edward R., and Jeanne S Poindexter. *Bacteria In Nature*. New York: Plenum Press, 1985.

[7]Foster, W. D. *A History of Medical Bacteriology and Immunology*. London: Heinemann Medical, 1970.

[8]Leadbetter, *Bacteria in Nature*, 3.

[9]Foster, *History of Medical Bacteriology and Immunology*, 5.

[10]Foster, *History of Medical Bacteriology and Immunology*, 9.
[11]Foster, *History of Medical Bacteriology and Immunology*, 4.
[12]Foster, *History of Medical Bacteriology and Immunology*, 6.
[13]Foster, *History of Medical Bacteriology and Immunology*, 6.

Homesickness at the Calgary Stampede

Rosalind Fleischer-Brown

When I was ten years old, I went to the Calgary Stampede for the first time. All I remember is watching a rodeo show and winning a stuffed frog. Visiting my family in Calgary and Edmonton was my favourite thing to do – spending hours at the waterpark in West Edmonton Mall, hiking in the mountains, having endless barbeques and going to Dairy Queen before every flight back home. I grew up in the Czech Republic with my dad, who's from Alberta, so we visited our family every few summers.

Somewhere during my late teens, I decided to move to Edmonton for university, foolishly thinking that moving to Canada will be as fun as going to the stampede when I was ten. I was wrong.

Ten years later, I was standing in the middle of a concert crowd at the Calgary Stampede, fighting off an anxiety attack. A few months after I moved to Edmonton, I was visiting my family in Calgary. My cousin was working at the stampede, so I went to pick her up. Arriving a bit early, I wandered to the closest stage. I think it was Nickelback, but I don't quite remember, because all I saw was a sea of cowboy hats, headbanging to songs I didn't know. Where the hell did I move to? I kept thinking. I felt so displaced, dizzy and homesick.

I pushed my way out of the crowd and walked into some random hallway near where my cousin was finishing her shift. There was a tiny souvenir store, so I wandered in. In classic Calgary Stampede fashion, every item had a cowboy on it and some tacky logo. The store had boots, hats, pillows, key chains, pencils, blankets, and God knows what else. I don't remember much, besides the overwhelming sense of panic, dissociation and nausea. I just wanted to be home. I left the store, leaned against the wall and slid down to the floor.

"Homesickness is like most sicknesses. Eventually it moves on to someone else," says a priest to a young Irish girl, after she moves to New York in the movie Brooklyn (2016). I saw this movie a month after I moved, and felt so understood, comforted and hopeful. Two years later, when I was re-watching the movie, I still empathized with the protagonist, but I was no longer in her place, adjusting to a new home and creating a new life. A few years later, when I was working at a restaurant, I was chatting to my new co-worker, who had just moved to Edmonton from Berlin, and I realized that the priest in that movie was right: homesickness passes and moves on to someone else.

I left for Prague a year ago to be with my closest family during the covid-19 pandemic. A month ago, I watched Brooklyn again to commemorate my fifth anniversary of moving to Canada and I realized I was homesick for Edmonton. Edmonton was difficult to love at first; I hated shoveling, the winters, the overly friendly attitude that seemed fake, and I hated the city itself. But over the years, I fell strangely in love with it. I made friends and lost friends; I moved houses; I lost jobs and found new ones. The more boring and stable Edmonton felt, the more I knew I was at home, and now I would do anything to snap my fingers and find myself back in that silly city with ugly buildings and freezing winters.

Biography

Rosalind Fleischer-Brown is a fifth-year psychology and English student at the University of Alberta. She published two books for the Antarctic Institute of Canada and translated a children's book from English to Czech. She also published a literary essay in the Glass Buffalo and published several articles in The Gateway, the University of Alberta's magazine.

Faculty of Arts, fleische@ualberta.ca

Austin A. Mardon is a fellow(hon) of the Royal Society of Canada and a member of the Order of Canada. In his youth he was part of a NASA expedition near the South Pole recovering meteorites.

Catherine A. Mardon is a Dame Commander of the Papal Order of St Sylvester and with her husband a ten-minute audience with Pope Francis. She is a retired attorney.

John DosSetor Health Ethics Centre, Faculty of Medicine, dossetor.centre@ualberta.ca

Dendrochronology – An Overview

Patricia D'souza, Dr. Auston Mardon

Quaternary is the current most recent of the three periods of the Cenozoic era. To research this period Dendrochronology is an important dating method used by many scholars. It is a science that dates annual tree rings to their exact year of formation and provides us with the environmental condition of that time.

Dendrochronology enables us to figure out sunspot cycles through tree rings. The link between tree-ring widths and climate gives us an insight into paleoclimate. It also introduced us to cross-dating, a technique used to match tree-ring widths from one tree to corresponding patterns for the same years from another tree (Panyushkina, 2011). Tree-ring analysis or dendrochronology is studied based on trees in temperate to boreal regions and austral regions to grow one growth increment of xylem cells per year on the outer portion of the wood stem, underneath the bark (Jacoby, 2000). Due to the cycle of xylogenesis during the growing season, deviations are generally caused by annual density fluctuations in tree rings. This definition is based under normal conditions where tissue growth activity stops once a year. Each tree ring can thus be dated to one calendar year, which is one of the main principles of dendrochronology (Micco, 2016). Dendrochronology can also be subcategorized into other fields, one of them being dendroclimatology. Dendroclimatology is also known as dendroecology and is essential for dating fossils as well as paleoecology. Tree rings derive from seasonal tissue growth activity, which makes them highly seasonal. Annual tree-ring width also relates to paleoclimate and atmospheric concentrations of carbon dioxide and oxygen, temperature changes, precipitation, and resembles climatic growth conditions (Lüttge, 2017). Dendrochronology is possible because many trees produce an annual growth that is easily identified by anatomical features that precisely show one year's growth from the next. These processes are driven by seasonal changes in climate that induce annual dormancy and allow for the reliably in the annual nature of growth rings. The chemical composition, size, and structure of an annual tree ring constitute a record of conditions in the tree's environment at the time the ring was being formed, hence a given parameter of that growth can be used as a proxy for environmental conditions at the time the ring was formed (Schneider, 2011). In temperate climates, endogenous, physiological, and age-dependent processes which are directed by mechanical stress, and locally variable light and water conditions have a greater influence on tree growth than regional climatic conditions (Schweingruber, 2007). Dendrochronological methods help with identifying tree age structures and can clarify the effects of changes in climate on tree establishment. Since the effects of tree mortality are often being inaccurately assessed, a variety of approaches have been proposed to deal with information loss in age structure analyses. At the same time, it is important to remember that dendroecological studies aiming at relating tree derived histories with past climate should consider the possibility that many plants might not survive to be recorded at the date of the study (Aráoz, 2015).

Tree-ring analysis provides precise dates of archaeological features, as well as the chronological sequence of disturbance impacts to surrounding woody vegetation and associated land-use practices. If the actual calendar years of a sequence of growth rings are known, other samples containing matching ring patterns can be reliably dated as well (Strachan, 2013). To achieve responsible research conducted on all levels and in all fields of science using specimens just like in dendrochronology proper preservation and organization is a must. The specimens need to be readily available for new analyses, confirmation or reinterpretation of results. Long-term preservation of the wood specimens and associated data are crucial to the field (Creasman, 2011). For example, in conifers, the most noticeable formation in a temperate tree ring is the earlywood-latewood succession. When the temperature is mild, soil water content is high, and the photoperiod is increasing, the light-coloured, low-density earlywood is the first part of the ring, formed at the beginning of the growing season. The darker, higher-density latewood forms during the second half part of the growing season (summer and early autumn). When the temperature is higher, soil water content is lower. (Rossi, 2014). Likewise, in the study of dendrochronology formulating fully regulated solutions for image-based tree ring detection is difficult. Although three main groups of wood-conifer, ring-porous and diffuse-porous anatomical structures

49

can be easily identified, the design of tree rings can sometimes greatly vary between tree species of the same group. Moreover, differences can sometimes be observed between tree rings representing one tree species based on the environment in which it grew, or even in one specimen. All these factors can be used to generate helpful wood images. Therefore, although accurate solutions that can be applied for all tree species are desired, most approaches concern only one tree species (Fabijańska, 2017). Similarly, the growth of any individual ring depends not only on the environmental conditions at a given time but also on mechanical stresses such as slope, snow creep, or partial wind-throw. Each of these conspires to displace the stem from the vertical. The response of a tree to such displacement is to develop "reaction wood" that produces a ring that is not uniform around the stem. The resulting rings become elliptical due to the greater growth on one side of the stem than the other owing to an increase in localized cambial activity. In the case of conifers, the greater growth occurring on the lower side of the stem is referred to as "compression wood." This growth is analogous to a buttress placed against a leaning wall. In deciduous trees, the same result is achieved through greater growth on the upper side of the stem, which is termed "tension wood" and is analogous to the stay on a mast or guy on a tent pole (Courtin, 2004).

Dendrochronology and its many branches can be used to solve many real-life problems such as model groundwater levels. Dendrohydrology is another branch of science that uses dendrochronology to investigate and reconstruct hydrologic processes, such as streamflow and past lake levels (Gholami, 2015). It is associated with streamflow through the common responses of tree-growth and streamflow to variations in net precipitation and runoff. The statistical relationship between time series of tree-ring and streamflow has been exploited for multi-century reconstructions of flow for river basins in many parts of the world. Dendrochronology has provided an effective and accurate evaluation of the magnitude and duration of severe hydrologic droughts (Meko, 2012). Moreover, it can be used to simulate other hydrological parameters such as streamflow discharge and climatological parameters (Gholami, 2015). Another subfield of Dendrochronology is dendrogeomorphology. It is used in places where snow avalanche tracks are often a common feature of sub-alpine forests, in these places the tracks prove to be a useful resource to examine the past avalanche activity. These records show the calendar dated tree ring records of the damage (Luckman,2010). Patterns in the rings related to climatic or disturbance factors can also be used to infer or reconstruct environmental conditions related to the historical activity (Strachan, 2013). In another study conducted by Babes-Bolyai University, they studied the snow avalanche activity in the Parâng Ski Area using tree rings. In this study, analysis was conducted by taking 11 stem discs and 31 increment cores were extracted from 22 spruces impacted by avalanches, to obtain more information about past avalanches activity. Using the dendrogeomorphological approach they found 13 avalanche events that occurred along Scărița avalanche path, from 1935 until 2012, nine of them produced in the last 20 years. The tree-rings data inferred an intense snow avalanche activity along the avalanche path (Meseşan, 2014). Dendrochronology methods do not only stop at avalanches but also help evaluate flood hazard assessment. In a study conducted by Ruiz-Villanueva in Central Spain, one of the main problems of flood hazard assessment in ungauged or poorly gauged basins is the lack of runoff data. To overcome this issue dendrogeomorphic time series can be used to reconstruct and compile information on 41 flash flood events since the end of the 19th century (2013).

Dendrochronology has many branches, one of them being dendroclimatology. In a study conducted by Gothenburg University, they looked at possible links to contemporary climate change, aerial photograph analysis and dendrochronology were accessed to combine them to study recent pine cover changes at Anebymossen, to do so a peat bog was raised in south-central Sweden. Further study was done by combining dendroclimatology and aerial photograph analysis, it is possible to get a detailed picture of long-term temporal and spatial tree cover changes on peatlands, as well as a possibility to date changes in growth rates with annual resolution and relate them to climate or anthropogenic changes (Linderholm, 2004). Overall, it can be concluded that dendrochronology is a useful method that can be used to date and give us an in-depth investigation history.

Works Cited

Alfaro, R., Campbell, E., Hawkes, B., Pacific Forestry Centre, Mountain Pine Beetle Initiative, Canadian Electronic Library, & Canada. Natural Resources Canada. (2012). *Historical frequency, intensity and extent of mountain pine beetle disturbance in British Columbia* (Mountain Pine Beetle Initiative working paper; 2009-30). Victoria, B.C.: Natural Resources Canada.

Aráoz, E., & Morales, J. (2015). Modeling unobserved variables in dendrochronological age structures improves inferences about population dynamics. *Canadian Journal of Forest Research, 45*(12), 1720-1727.

Creasman, P. (2011). Basic principles and methods of dendrochronological specimen curation. *Tree-Ring Research, 67*(2), 103-115.

Courtin, G., Fairgrieve, S., Fairgrieve, SI., American Society for Testing Materials, & ASTM International. (2004). *Estimation of Postmortem Interval (PMI) as Revealed Through the Analysis of Annual Growth in Woody Tissue.*

De Micco, V., Campelo, De Luis, Bräuning, Grabner, Battipaglia, & Cherubini. (2016). INTRA-ANNUAL DENSITY FLUCTUATIONS IN TREE RINGS: HOW, WHEN, WHERE, AND WHY? *IAWA Journal, 37*(2), 232-259.

Fabijańska, A., Danek, M., Barniak, J., & Piórkowski, A. (2017). Towards automatic tree rings detection in images of scanned wood samples. *Computers and Electronics in Agriculture, 140,* 279-289.

F. Meseşan, O. Pop, & Ionela Georgiana Gavrilă. (2014). SNOW AVALANCHE ACTIVITY IN PARÂNG SKI AREA REVEALED BY TREE-RINGS. *Studia Universitatis Babeş-Bolyai: Geographia, LIX*(2), 47-56.

Gholami, V., Chau, K., Fadaee, F., Torkaman, J., & Ghaffari, A. (2015). Modeling of groundwater level fluctuations using dendrochronology in alluvial aquifers. Journal of Hydrology, 529(3), 1060-1069.

Helama, S., Hyttinen, O., & Salonen, V. (2012). Late Weichselian varve archives re-explored to assess proglacial sedimentary chronologies using the principles of tree-ring analysis. *Progress in Physical Geography*, 36(2), 187-208.

Jacoby, G., Noller, Jay Stratton, Sowers, Janet M., & Lettis, William R. (2000). Dendrochronology. AGU Reference Shelf, 11-20.

Luckman, B., Stoffel, M., Bollschweiler, M., & Butler, D. (2010). Dendrogeomorphology and Snow Avalanche Research. In *Tree Rings and Natural Hazards: A State-of-Art* (Vol. 41, Advances in Global Change Research, pp. 27-34). Dordrecht: Springer Netherlands.

Lüttge, U. (2017). From dendrochronology and dendroclimatology to dendrobiochemistry. *Trees*, 31(6), 1743-1744.

Linderholm, H., & Leine, W. (2004). An assessment of twentieth century tree-cover changes on a southern Swedish peatland combining dendrochronoloy and aerial photograph analysis. *Wetlands*, 24(2), 357-363.

Meko, D., Woodhouse, C., & Morino, K. (2012). Dendrochronology and links to streamflow. *Journal of Hydrology*, 412-413, 200-209.

Panyushkina, Irina. (2011). Dendrochronology. https://www. researchgate.net/publication/259220539_Dendrochronology

Ruiz-Villanueva, V., Díez-Herrero, A., Bodoque, J., Ballesteros Cánovas, J., & Stoffel, M. (2013). Characterisation of flash floods in small ungauged mountain basins of Central Spain using an integrated approach. *Catena*, 110(C), 32-43.

Schweingruber, F., & SpringerLink. (2007). *Wood Structure and Environment* (Springer Series in Wood Science).

Rossi, J., Nardin, M., Godefroid, M., Ruiz-Diaz, M., Sergent, A., Martinez-Meier, A., . . . Thioulouse, J. (2014). Dissecting the Space-Time Structure of Tree-Ring Datasets Using the Partial Triadic Analysis. *PLoS ONE*, 9(9), E108332.

Schneider, S., Root, T., & Mastrandrea, M. (2011). Dendrochronology. *Encyclopedia of Climate and Weather*, Encyclopedia of Climate and Weather.

Strachan, Scotty, Franco Biondi, Susan G. Lindstrom, Robert Mcqueen, Peter E. Wigand, and S. Lindstroem. "Application of Dendrochronology to Historical Charcoal-Production Sites in the Great Basin, United States." *Historical Archaeology* 47.4 (2013): 103-19. Web.

St. George, S., Meko, D., & Cook, E. (2010). The seasonality of precipitation signals embedded within the North American Drought Atlas. The Holocene, 20(6), 983-988.

The Lockdown – A Story

Rosalind Fleischer-Brown

"I won't be able to take your offer due to COVID reasons," I replied to the manager of a not-for-profit organization. I applied for a social worker job, where I would be working with people struggling with addiction and the risk of getting infected with corona virus is very high. Because I live with my mother, who is uncomfortable with such a high probability of risk, I had to turn the job down. I guess back to working from home it is.

A few days later, I chat with a friend of my dad's about a possible internship. She is a play therapist and since I want to be an expressive arts therapist, I figured she might be a good person to talk to. However, due to the current situation, she barely works and has little to offer me, besides a few contacts, an apology, and empathy for my frustration.

As a young adult in my mid-20s, finishing university and having more time, energy and skills than ever before, I feel incredibly frustrated with the situation: instead of taking on interesting jobs, internships, travelling, and volunteering, I am forced to stay inside and find something to do. I have roughly six months of lockdown experience, living in one of the worst countries in the world, in terms of the daily cases of infected people.

At the end of March 2020, I took a repatriation flight from Montreal to Prague, where I grew up, to spend time with my family, especially with my grandmother who was battling cancer at the time. (Fortunately, she recovered from it!) I planned to stay a few months only, but finding out that my university classes will be online, I decided to stay in Prague, having no reason to go back to Canada. I have been here for nearly a year and I must say that being in lockdown in this country is nothing short of excruciating. The Czech Republic has handled the pandemic absurdly poorly, leaving us in a strict lockdown since October of last year.

The covid-19 pandemic and its consequences has affected everyone on some level, but the increase in experiencing negative emotions is jarring. Psychologists and other workers in the mental health field are in demand more than ever. Therapists say that the world is experiencing collective trauma. It seems that we are all sick of this pandemic; individually and globally. Living in our tiny Zoom screens, communicating scarcely with other human beings face-to-face and constantly checking the news, only to find very little has changed, this pandemic is bound to create some discomfort. Lack of motivation, restlessness, boredom, loneliness, frustration and anxiety are just a few emotions I have heard my friends and family struggle with. Personally, restlessness, frustration and loneliness seem to haunt me the most.

More than once have I heard someone refer to these days as "Groundhog Day". If you haven't seen this classic 90s comedy, Bill Murray wakes up one day, having to relive the same day over and over again. And lockdown is starting to feel like this: every day seems the same, because we are all forced to stay inside and unless we are able to create nuance in our day-to-day life, our days become quite similar. This can create a sense of boredom, restlessness, dejection and other feelings. I wake up every day, go for a run (usually the same route), shower, do my homework, attend classes (over Zoom, of course), watch some Netflix and go to sleep. The next day looks the same. Occasionally, I do yoga, play guitar and sometimes I read. Other times, I video chat with friends in Canada. Either way, my days are very similar and sometimes I wake up feeling so down, I go back to sleep, just so that I can avoid the monotony. The obvious solution to this problem would be to create some novelty in my days, so I've ordered some yarn and decided to knit a scarf for myself.

But in all seriousness, if you are struggling to find some joy, excitement or motivation in your day, try and find something new to do. Is there any skill you have always wanted to learn? A book you've been meaning to read for a while; or a book you've been meaning to write? Is there a job, project or internship you can do from the comfort of your home? If you've tried everything and nothing really works, do something completely extraordinary. A few months back, my friend and I skateboarded to a castle nearby, climbed to the window of it, which was elevated on a tiny, rocky hill, and broke into a castle. Was it scary, reckless and stupid? Yes. Did this experience enrich my lockdown days? Absolutely. I had an experience unique to that particular windy Thursday that I will remember fondly. All the other lockdown days blur together. If there is any time to try new things, learn new skills and pick up new hobbies, it is now. We will never have this much time on our hands again.

As a consequence of this pandemic the world is changing and some say online learning and working from home may be the norm for the next few years, if not permanently. As depressing as this is, there is little we can do about this fact, except to adapt. Personally, I am still holding onto hope that I will be able to do an in-person master's degree in 2022, but if the option is not available, I will have to do it from home. So, while I am sitting in lockdown with my Bachelor of Arts, looking for jobs, I may as well engage in some online learning and workshops, even if I won't be doing a master's, because if the pandemic has brought any positives, it is an increase in online education that is accessible to all. Talks, workshops, classes and seminars are a big hit right now. So, if you have nothing else to do and you want to pump up your resume, or learn for the sake of educating yourself, browse Facebook and you will find an online event very quickly.

Next to struggling to find a job, the consequences of social isolation are negatively affecting our lives. Having connection with others is crucial, because, at the end of the day, you may still be struggling to find a job or have to process other consequences of this pandemic, and this is easier to do when you have someone to talk to. Some of us live with roommates, family members or partners and some of us live alone. Some of us are surrounded by others and feel lonely, while others are perfectly content on their own. Regardless of what our personal situation may be, human connection is crucial to our well-being. People are social animals and forcing them to stay couped up inside and avoid others goes against human instinct to socialize. Friend hangouts turned into video chats, phone calls and texts, as if we did not have enough screen time working or watching lectures. Personally, I enjoy sending voice memos to my friends in Canada and getting their voice memos in reply. It's much more intimate than sending texts, because you get to hear the other person's voice.

While we cannot hug each other, go on dinner dates or simply grab coffee with someone (at least not here in the Czech Republic), socially distanced walks are better than having no social interaction. If that isn't possible, a phone call or video chat works well, especially if you want to reach out to your grandparents, who maybe struggling with loneliness, anxiety and fear more than any of us. Check on them and see how they are doing; if they are vaccinated, you may be able to give them a hug sooner than you think.

Being in lockdown presents immense challenges, both socially, emotionally and practically. While this reality is difficult to navigate and can encourage dejection, hopelessness, depression and loneliness, this situation – unfortunately – is outside of our control and the only thing we can control is our reaction to it. So, even though it sucks to work from home, do a degree online, or be unable to hug your friends, do the best you can to mitigate this crisis. While it is unpleasant, it will not last forever; maybe just the next six months to a year.

Biography

Rosalind Fleischer-Brown is a fifth-year psychology and English student at the University of Alberta. She published two books for the Antarctic Institute of Canada and translated a children's book from English to Czech. She also published a literary essay in the Glass Buffalo and published several articles in The Gateway, the University of Alberta's magazine. She is currently living in Prague with her family.

Faculty of Arts, fleische@ualberta.ca

Austin A. Mardon is a fellow(hon) of the Royal Society of Canada and a member of the Order of Canada. In his youth he was part of a NASA expedition near the South Pole recovering meteorites.

Catherine A. Mardon is a Dame Commander of the Papal Order of St Sylvester and with her husband a ten-minute audience with Pope Francis. She is a retired attorney.

John Dossitor Health Ethics Centre, Faculty of Medicine, dossetor. centre@ualberta.ca

Living with Mental Illnesses as a South Asian Youth

Fatima Saleem

I am a mental health advocate, and I've written this story based on a conversation with a friend who deals with mental illness on a continual basis but wishes to remain anonymous; he suffers from both attentional deficit hyperactivity disorder (ADHD) and depression. Notably, the conversation highlighted the stigma surrounding mental health and how it is an issue in modern society. This statement is particularly true for men and immigrant communities:

We all lie. Sometimes, these lies become solidified in our lives to the extent that we start believing that we are who we pretend to be. For me, I pretended to be a person with control who didn't need help and had the strength to conquer every task all by themselves. My facade was a brick wall built without mortar, bound to fall down. I tried to stop my thoughts, but it felt impossible to maintain a perfect image. Living with mental illnesses was like forever being caged in a dark escape room, without handles on the doors, and without any clues or hints on how to escape.

Closing my eyes and listening to music was my escape from reality. When thinking of a support system, I immediately turned to the lyrics, "Solo ride until I die" by Bebe Rexha. The song, "Me, myself and I" is the epitome of what getting help felt like to me as I struggled to control my anxiety. As I grew older and faced even more challenges, from the boy who bullied me in class to my parents ignoring me, I acted as if these were trials that I had to face alone to grow. A test that I was thrown in and had failed to succeed at. I didn't always understand what was going on in my mind. I vividly remember sleepless nights, turning back and forth in my bed while gazing at the soulless ceiling, trying to solve my problems alone from fear of being judged and labelled. Without even realizing it, I had created a bubble filled with my negative thoughts with words like "incompetent" and "shame" that would constantly bounce around me, encompassing me in depression. My days were filled with attempts to distract myself with gaming, separating myself from reality by delving into fantasy land, where I could control the actions and outcomes of the characters. I didn't want to be seen as someone who couldn't take care of themselves. If a family member walked into my room, I would forcefully remind them that my problems were mine alone, not theirs. Gradually, my school attendance got worse and as a result, my grades suffered which pushed me into a cycle of feeling worthless and insignificant. I knew that I couldn't let others see this side of me, so I put on a pretense smile, masking hidden layers of my inner turmoil and suffering. Fear of stigma and judgement was always lurking in the shadows. But, would you blame me for having this viewpoint given the culture I was raised in?

As an immigrant who grew up in the South Asian community, it felt like support was never available. My conservative parents attributed my emotional outbursts to problems relating to my physical health, as if my feelings didn't garner attention. No one tried to dig deeper into the reasons why I experienced mood swings – the experiences and events that I had gone through, or even bothered to ask me how I was feeling. Even in passing conversations, it was considered taboo to discuss my emotions, sad, angry, embarrassed, whether in public or private. If you listen to a typical conversation of a South Asian community, you won't hear even a single word about mental health. The wide-spread belief was that "boys don't cry." I continued pretending to be strong, finding it difficult to talk to my parents and friends who would likely have labelled me as "weak." I started to internalize the stigma surrounding mental health, feeling ashamed for who I was, and thus, the need for a facade of independence and false strength was created. I felt hopeless.

My emotions were like my baby sister: you couldn't predict when she would cry, feel agitated or lose control. One mindless mundane morning, I experienced an emotional outbreak from all the feelings I had bottled up, refusing to leave my bedroom for days and screaming for hours at a time. The outside world was too overwhelming. It felt like there were shackles holding me down, preventing me from seeking the clues necessary to leave this make-shift escape room. There was no purpose in getting out of bed because I felt that I didn't belong in the world.

When I looked in the mirror, I realized I couldn't put the mask back together. I could no longer pretend to have my life together, but this change in attitude didn't take days or months as I had initially expected, it took years. Alongside my own acceptance, my parents had started noticing my mental health struggles and its causes, and in turn, sought out support for me, including therapy and listening to my concerns. Then, it struck me - I didn't have to face my struggles alone. It was okay to ask for help. For the first time, I created a plan to take charge of my life by seeking support and discarding the pretense of being societally strong. I now know that opening up was the strongest thing I could have done – in fact, seeking help is a sign of strength. I was not alone, and neither are you.

Author Biography:

Biomedical sciences student by day, writer by night, Fatima Saleem is a young Pakistani-Canadian author based in Calgary, Alberta with a passion for mental health education. When she is not writing at her favorite coffee shop, Fatima can be found advocating for mental health support in minority communities through volunteering at the Kids Help Phone and spearheading the Art of Recovery Club at her university. She aspires to use literature to eliminate the stigma of mental illness, teach empathy and showcase behind-the-scenes mental health struggles of people of color, particularly youth in the South Asian community.

Pestilence, Plague Doctors, and Pandemics: An Overview of Historical Pandemic Response in the Western World

Jessica C. Henschel

Disease has been a constant enemy of humankind throughout history and continues to be a major social and economic challenge for societies. As we continue to manage the COVID-19 pandemic in the present times, it is beneficial to look back upon previous instances of pandemics to assess how humans have reacted to viruses in the past. The approaches and responses to plague have varied significantly in the western world, shaped by the lack of knowledge and understanding of bacteria and viruses. This paper overviews the two most significant pandemics in the west prior to the 20th century and analyses how the doctors and general public responded to a seemingly invisible threat.

As the earliest recorded pandemic in the western world, the 430 BCE Plague of Athens offers crucial insight into the way ancient citizens responded to mass disease and death. The plague was recorded to have devastated the city in more ways than just population. Primary sources of the plague are based on the accounts of Greek historian and general Thucydides, thus allowing for a detailed first-hand account of the death and turmoil, and ramifications (Wycombe-Gomme, 2020). Thucydides' (c. 460-400 BCE) primary work, History of the Peloponnesian War, is one of the only primary sources that offers an eyewitness account of both the plague and the events of the Peloponnesian War. After catching and surviving the plague, he was elected as general and given a fleet. However, after failing to prevent the capture of the major Athenian city Amphipolis by the Spartans, he was sentenced to exile. It was in exile that he was able to write his account of the pandemic and the war. Presenting a comprehensive description of the symptoms — high fever, rash, and diarrhoea — and effects of the plague, Thucydides provides a narrative about the Athenian society during this time (Littman, 2009).

When the plague entered Athens in 430 BCE, the city state was under siege by Sparta during what is considered the second part of the Peloponnesian War (431-404 BCE) (Littman, 2009). Led by the famous statesman Pericles, the Athenians were weakened from fighting for nearly 3 decades. Based on Thucydides' writings, the plague was thought to have originated in Ethiopia and traveled through Egypt and the Mediterranean to arrive through Piraeus, the port of Athens. In the next 3 years, between 75,000 and 100,000 people died — over 25% of Athens' total population. Based on Horgan's (2016) research, many of the infected died 7-9 days after the first appearance of the disease, quickly decimating the numbers of workers and most importantly, soldiers. Thucydides recalls those who did survive the illness suffered from blindness, memory loss, and permanent disfiguration.

Initially the plague was thought to have been caused by the poisoning of the water sources by the Peloponnesian enemies (Martínez, 2017). This theory was later discarded when the sickness moved inside the city walls, where the water wells were not connected to the Piraeus. Blame then turned towards the gods, and the people were convinced they were being punished due to belief that the god Apollo (often associated with pestilence) had sided with the Spartans (Martínez, 2017). In modern times, there are approximately 30 different diagnoses of the pathogen that caused the plague, but researchers have narrowed it down to either smallpox or typhus (Littman, 2009).

Aside from the horrendous number of deaths, the ramifications of the plague were apparent in many different spheres of society. Specifically, the movements for religious secularization that resulted from general lack of faith, as well as the restriction of women's rights in society and in the home (Martínez, 2017). However, the impacts of the plague reached beyond the social structures of Ancient Greece, influencing the outcome for the Athenians in the Peloponnesian War — a crippling defeat. The Spartans' victory meant unbearable economic consequences for Athens, ones in which their declining city would not be able to support in the future.

Similar to the Plague of Athens, various societal and political shifts occurred during the time of the infamous Black Death in Europe. Between 1347-1351, the epidemic commonly known as the Black Death swept across Europe, killing more than 20 million people (25% of the population) (Lindemann, 2010). Unlike in Athens, there were many sources and first-hand accounts in which information was recorded about the plague, informing us of the atmosphere in Europe when the disease first appeared. As Lindemann (2010) details, these sources informed us that the plague was believed to have entered the continent from 12 ships from the Black Sea docked at the Sicilian port of Messina. Most sailors aboard the ship were either dead or severely ill and covered in black boils that oozed blood and pus. Despite the Sicilian authorities' attempt to rid the harbor of all the so called "death ships", the plague had already begun to spread (History.com, 2020).

The Black Death devastated Europe and Asia during the first 5 years after its arrival in Messina but continued to last for 400 years after until it finally vanished (Lindemann, 2010). It was thought to be spread by trading ships, making it rampant and infectious to every major city with a port. It was later hypothesized that the true carrier of the plague on these trading ships was rats, as they were carriers of the bacteria. Scientists and historians in modern times have suggested that the rats were carrying the Yersinia Pestis bacteria (bubonic plague), which would be hosted by fleas on the rodents (Lindemann, 2010). Once the carrier rat died, the fleas would hop onto humans and bite them, thus infecting them with the bacteria. Once a human was bitten, the site swelled to form a painful and large bubo, most often in the groin, thigh, armpit, or neck. Victims began to show signs 3-5 days after they were initially infected, usually dying another 3-5 days later (80% of the cases) (Benedictow, 2005).

Despite many advertisements for remedies and healers that could cure the plague, mass death still occurred late into 1351 and beyond (Lindemann, 2010). As no medicines or doctors seemed to help, many came up with alternative methods of explaining and living through the plague. There were various explanations and responses to the Black Death, with the most prominent being religion. The common understanding in the predominantly Christian areas was that God was punishing the sinners with the great dying and saving the believers (Lindemann, 2010). The churches responded by sending out prayers to be recited by the public and held large masses to pray for reprieve. Due to this belief, they thought that the only way to be saved was to gain God's forgiveness and purge their communities of heretics (History. com, 2020). Another well-known response to the plague were a group of individuals called flagellants. Upperclassmen were known to join processions of traveling flagellants, who engaged in public displays of punishment (History.com, 2020). They would beat themselves and others in with leather straps studded with metal in public areas for 33 and ½ days. Townspeople would watch as these men inflicted pain on one another 3 times a day. This movement was targeted towards repenting for their sins and seeking penance.

As Lindemann (2010) emphasizes in her review of medicine and early European society, it is still unknown how great of an impact the Black Death had on European society as a whole. People began to migrate to less populated areas, resulting in clearing of forests and reliance on animal husbandry (Benedictow, 2005). Small-scale trade was forced to increase due to the restrictions on naval trade, resulting in economic losses. However, one of the most significant societal and scientific changes that occurred due to the plague was the increase in public health measures. The loss of life was significant and medical and public health orders began to develop out of pure necessity to limit the spread (Lindemann, 2010). Officials began to set up quarantines, where the infamous plague doctors in their beak-like masks would visit to assess symptoms. Subsequently, fumigation was arranged for people in quarantine and for their possessions in an attempt to clear the miasma or "bad air", which was how the plague was thought to be spread (Lindemann, 2010). Additionally, special hospitals were built for the further isolation of patients and large gatherings were prohibited.

Looking back at the impacts of the Plague of Athens and the Black Death, many parallels can be drawn to COVID-19 and our current situations. Although the responses to mass illness were met with religious sentiments rather than science, the same fear and panic befell these cities as we are facing today. In hindsight, we have these earlier societies to thank for the development of our well-informed and scientifically driven public health orders that have saved millions of people. If not for our modern understanding of bacteria and infection, COVID-19 may have been as deadly as the plagues across Europe in the 14th century and earlier.

References

Benedictow, O.J. (2005, March 3). The black death: The greatest catastrophe ever. History Today. https://www.historytoday.com/archive/black-death-greatest-catastrophe-ever

History.com. (2020, July 6). *Black death.* https://www.history.com/topics/middle-ages/black-death

Horgan, J. (2016, August 24). *The plague of Athens 430-427 BCE.* Ancient History Encyclopedia. https://www.ancient.eu/article/939/the-plague-at-athens-430-427-bce/

Lindemann, M. (2010). *Medicine and society in early modern Europe* (2nd ed.). Cambridge University Press.

Littman, R.J. (2009). The plague of Athens: Epidemiology and paleopathology. *Mount Sinai Journal of Medicine,* 76(5), 456-467. http://doi.org/10.1002/msj.20137

Martínez, J. (2017). Political consequences of the plague of Athens. *Graeco-Latina Brunensia* 22(1), 135-146. http://doi.org/10.5817/GLB2017-1-12

Wycombe Gomme, A. (2020, February 6). *Thucydides.* Encyclopedia Britanica. https://www.britannica.com/biography/Thucydides-Greek-historian

Repatriating Back Home During the COVID-19 Pandemic

Rosalind Fleischer-Brown

"There's a repatriation flight for all Czech citizens from Montreal this Sunday. You should call the embassy as soon as possible. Here's the link!" my mom messages me on Wednesday, the 25th of March, 2020. Barely awake, I scan through the article, grab my phone and call the Czech embassy in Ottawa.

"We need to know if you are interested right now," the woman says.

"Yes, I am!" I blurt out, unsure of how I am going to get myself to Montreal before Sunday.

"Great, just send us your contact information and passport number."

I spend the next two hours furiously booking a flight, an Airbnb, alerting my professors, landlord, roommate, family and friends in Edmonton about the change. By Friday, I find myself in a comfortable little Airbnb in Montreal. My determination to pretend I am on vacation is ruined by heavy rain and the sudden realization I have to avoid people and wash my hands constantly. On Sunday, I board a flight full of Czechs and Slovaks ready to go home, after their work holiday in British Columbia or something. I return home to see my grandmother, who is battling cancer and because I want to be with my immediate family during this pandemic. Not to mention that my classes moved online, I was laid off work and I had a challenging roommate I did not want to live with anymore.

I thought I would be back in Edmonton in July to resign my lease, find a job and prepare for another semester of university. I packed for three weeks, planning to be in Edmonton after three months, but turned out that the University of Alberta did not re-open, the pandemic worsened, and I had no reason to return to Edmonton, so, yes: I am still here. It has been exactly a year, since I repatriated back home.

Taking a repatriation flight in the midst of a global pandemic is a unique experience. Any other time, I have flown back to Prague, I felt excited, my suitcase filled with Christmas presents or summer dresses. This time, I was recoiled in a face mask, protecting my hands with gloves and squirting hand sanitizer on myself every fifteen minutes. Most people were supportive of my journey, but some judged me and made me feel guilty for travelling at this time, but I had to go. I had to see my grandmother. (If you are wondering, she beat cancer and I might see her soon, because she just got vaccinated!)

Living back home has been a strange experience. When I first moved to Edmonton, I was homesick for Prague for months, if not a year. Living abroad was extremely difficult and repatriating allowed me to feel comfortable, safe and secure again, surrounded by my closest family and friends, and I am grateful for this time. However, I built a life in Edmonton and leaving it so abruptly, I find myself grieving my old house, my old life, and everything that came with it. So, if repatriating taught me anything, it is that home is where the heart is and I seem to have left mine in Edmonton.

Biography

Rosalind Fleischer-Brown is a fifth-year psychology and English student at the University of Alberta. She published two books for the Antarctic Institute of Canada and translated a children's book from English to Czech. She also published a literary essay in the Glass Buffalo and published several articles in The Gateway, the University of Alberta's magazine.

Faculty of Arts, fleische@ualberta.ca

Austin A. Mardon is a fellow(hon) of the Royal Society of Canada and a member of the Order of Canada. In his youth he was part of a NASA expedition near the South Pole recovering meteorites.

Catherine A. Mardon is a Dame Commander of the Papal Order of St Sylvester and with her husband a ten-minute audience with Pope Francis. She is a retired attorney.

Taking Care of Your Mental Health Amidst the Pandemic

Parmpreet Kang, Austin Mardon

Parmpreet Kang

Parmpreet Kang is an Honours Life Sciences student at McMaster University. She is passionate in writing about important world issues and various fields she is interested in like biology, psychology, etc.

Austin Mardon

Austin Mardon, PhD, CM, FRSC, is an author and advocate for mental health. He is an assistant adjunct professor at the University of Alberta in the John Dossetor Health Ethics Centre and in the Department of Psychiatry.

What is one thing that everyone can say impacted them this year? COVID-19. A couple of years prior no one would have thought that we would be so ruthlessly impacted by a virus- a virus that has infamously shaken up the whole world and caused severe economic downfall, drastic changes to people's day-to-day lives, and countless deaths. Now, over a year later, the virus continues to pose a health risk, and people are encouraged to continue physical distancing and self-isolating. Although self-isolation is greatly important in minimizing your risk of contracting the virus, feelings of prolonged loneliness, boredom and stress are proven to have a serious toll on one's mental health in the long run. Hence, it's important to take care of your mental health. Here are some great tips to help you!

Disconnect from the News and Meditate:

Constantly hearing about devastation occurring around the world and obsessively monitoring the number of confirmed cases in one's respective area is not good for your mental health. Although it's important to stay informed, it's best to stick to reliable news sources and not overload yourself with news as this can make you feel overwhelmed and hopeless. Instead, spend more time disconnecting from your electronics and practicing mindful meditation and breathing exercises as a form of relaxation. If you are new to meditation, many resources are available to help you, from YouTube videos to popular phone apps like Headspace.

Stay connected with your loved ones:

It seems that a lot of people are mixing up physical distancing with social distancing. Although it is best to "physically distance" yourself from others, it does not mean you should "socially distance" and cut off all forms of communication with your friends and family. Talk to them, discuss your feelings, and keep each other updated about what you've been up to, whether that be through Facetime, text, or call. There are several more unique ways to stay connected and do fun activities despite being far. For instance, apps like "Netflix Party" allow you to watch Netflix with your friends by synchronizing video playback and providing a chat option while you watch.

Invest in Yourself:

Now is the perfect time to focus on yourself and improve, considering you have endless resources on the internet. Try a free online university course from a website like Coursera or Udemy, develop a new skill from Youtube, or learn a language from Duolingo. Develop a new exercise routine and improve your physical health. Try out new, healthy recipes. Take care of your skin. The opportunities are truly endless.

Do what you've been delaying:

You know that thing you have wanted to do for so long? Whether it's that book you've been meaning to read, hobby you've been neglecting, home renovation project you've been delaying- because you were just too busy and didn't have the time for it? Well now is the perfect time to do just that! Not only will you stay occupied, but it will be satisfying to finally complete it and check it off your to-do list.

Reorganize your Living Space:

Amidst the pandemic, many people are constantly spending time in the same environment every day. However, constantly being in the same environment can get agitating and boring, so spend some extra time rearranging your living space! This is scientifically proven to uplift your mood. In fact, Dr. Carrie Barron states that "an impact on the environment, whether an imprint or a removal, lifts mood, provides concrete satisfaction, and instills a sense of effectiveness".

Be kind:

Naturally, when you do something kind for someone else, your brain releases dopamine, which promotes positive feelings and uplifts your mood. Hence, try to do something nice whenever you can, like help out your elderly neighbors by picking up their groceries when you're out on your next grocery trip, or give a larger tip the next time you're ordering food to thank the essential workers who continue to put themselves at risk amidst this pandemic. Don't forget to wear a mask whenever you go out and interact with others!

Explore the Outdoors:

Physical distancing does not mean you need to completely stay indoors 24/7. You can still explore your city while ensuring you are a safe distance away from others. Get away from the daily routine and explore what nature's got to offer by visiting the beach, local trails, waterfalls, etc.

Go to 'Social-Distancing Approved' Events:

On the topic of exploring your city, check out some of the events your city is holding that allow you to have fun while staying safe! For instance, take a trip to a drive-in movie theatre, drive-in safari, or drive-in concert. Drive-in entertainment is definitely a theme in 2021!

Reflect and Go Easy on Yourself:

Take time out of your day to journal down your thoughts and reflect. Recognize your goals, and evaluate how effectively you have been achieving them, but remember to not be so harsh on yourself if you have not achieved as much as you hoped. Going through a pandemic is rough, and it's okay if you don't meet all your expectations. Take it day by day.

Reach out to a Professional if Need Be:

It's a stressful time for everyone and it's important to take the necessary steps to protect yourself from the virus and take care of your mental health, as it's just as important as your physical health! If you feel like you need help, don't hesitate to contact your primary health provider, a mental health professional, or a suicide prevention helpline.

Cultural Production and Gender Inequality in Video Games

Patricia D'souza, Auston Mardon

Video games are a commodity utilized by players all over the world, especially in North America. It is a multi-billion-dollar industry that is only expected to grow (Taylor, 2015). In Canada and the United States, most households have at least one gaming device that they use regularly (Taylor, 2015; Cade and Gates, 2016). This has allowed video games to produce culture in many ways in North America. First, video games are not only targeting people who want home consoles, but also consumers who are busy and need something to play with while travelling to work, school, etc. (Taylor, 2015; Cade and Gates, 2016). This production of portable systems is important because it demonstrates how inclusive cultural consumption is becoming in the gaming industry. Developers want to produce games and game consoles that target on-the-go consumers because they are becoming a growing target audience in gaming society. Second, video games are being used as a tool in education and therapeutic environments (Hartley, 2014). This has allowed the gaming industry to produce devices that are helpful in our society. For example, in Project EVO – a therapeutic game designed to enhance cognitive functioning in children and teenagers with autism – the consumers would try to steer an alien down a river and touch animals as they appear on an iPad screen. After playing this game for a month, researchers were able to see improvements in memory and cognitive functioning amongst the consumers ("Autism Speaks..." 2012). Therapeutic and Educational gaming allow for the reproduction of culture, and act as a mediator between educators/therapists and their students/clients. As a result, this allows cultural exchange to occur between both parties, which results in better educational and therapeutic tools and techniques being developed over time.

Games are also an excellent way for people to connect and socialize. Video games help to create social bonds among groups, and create identities amongst consumers, depending on the types of games that they consume (Taylor, 2015). These social bonds can be seen on many different levels, which can include parents playing video games with their children, senior citizens learning how to play the Nintendo Wii with volunteers, and friends playing video games together, both locally and online (Cade and Gates, 2016). The social implications of gaming are significant because they help to bring different cultures and generations closer together. Despite the production of sub-cultures within gaming (such as RPG gamers, PC gamers, casual gamers, etc.), these consumers are all a part of one big community: the gaming community (Taylor, 2015). The gaming community has become more inclusive to women, but inequalities in the gaming world continue to persist, and we still have a long way to go in terms of seeing women on levelled playing fields: both as protagonists in games, and as players and developers (Johnson, 2013).

Many of the earliest video games of the 1980s that featured a central protagonist deemed female characters as plot objects: housewives, maids, and damsels in distress that needed to be rescued by a more dominant male performer, such as a knight, spy, or an Italian plumber (Cuddy, 2008). Indeed, there were heroines in these 8-bit adventures that defied the norms of the stereotypical female protagonist, the majority of whom were thin with long legs, large breasts, and a feminine disposition; unfortunately, these nonstandard heroines were usually made to be ridiculed (Hardcore Gaming 101; Behm-Morawitz and Mastro, 2009). For example, in Mermaid Madness, a 1986 platformer, Myrtle is an overweight mermaid that is portrayed as an 'ugly ducking' not only to her crush (a diver), but to the player as well (Hardcore Gaming 101). The fact that female characters were perpetuated in this manner demonstrates the reality of a complete underrepresentation of strong, empowered female protagonists in 1980s and 1990s video games. Today, leading ladies are significantly more confident and independent than their earlier counterparts, yet many of them are still presented in a sexualized manner (Milford, 2016). To combat this, the best approaches for empowering women would be to diversify the roles that they play in video games, such as being a scientist, hipster, or athlete, and not emphasizing the protagonist's gender (Cade and Gates, 2016; Martins et al., 2009). This will ultimately bring gamers one step closer to appreciating and embracing female protagonists as powerful and compelling characters that bring an edge to an otherwise male-dominated culture.

Female protagonists are not the only ones to endure criticism within video game culture. Female players are also being scrutinized by male gamers based on the beliefs that women are less skilled at gaming than their male counterparts, and that women lower the quality of gameplay by working to promote inclusivity within the gaming industry (Milford, 2016). This not only represents how women are inferior in the field of gaming, but it also reinforces a power relationship and a group identity since men dominate the video game industry (Cate and Gates, 2016). The gender ratio of game developers in the industry is disproportionately skewed in favour of men over women. On the other hand, women that are working as game developers are significantly more likely to be targeted and harassed, particularly online, where Internet trolls can remain anonymous (Johnson, 2013; Milford, 2016). This online anonymity demonstrates how virtual gender discrimination tends to be overlooked as a serious issue, and how the patriarchy is normalized in video game culture, especially on the Internet (Behm-Morawitz and Mastro, 2009; Johnson, 2013; Milford, 2016). We cannot overlook the issues that women face in the gaming industry, both online and offline. People need to realize that the misogyny that women face in the world of video games is unacceptable, and that solving these systemic issues within gaming is crucial if we want to move forward in the gaming industry (Milford, 2016; Johnson, 2013).

References

Autism Speaks' DELSIA Funds Clinical Trial of Therapeutic Video Game. (2012, July 24). Retrieved January 30, 2018 from https://www.autismspeaks.org/science/science-news/autism-speaks%E2%80%99-delsia-funds-clinical-trial-therapeutic-video-game

Behm-Morawitz, E., & Mastro, D. (2009). The Effects of the Sexualization of Female Video Game Characters on Gender Stereotyping and Female Self-Concept. Sex Roles, 61(11-12), 808-823. doi:10.1007/s11199-009-9683-8

Cade, R., & Gates, J. (2016). Gamers and Video Game Culture. The Family Journal, 25(1), 70-75. doi:10.1177/1066480716679809

Cuddy, L. (2008). The Legend of Zelda and Philosophy: I Link Therefore
 I Am. Chicago, IL: Open Court.

Hardcore Gaming 101. 1980s video game heroines. Retrieved January
 29,2018fromhttp://hg101.kontek.net/inventories/80sheroines2.
 htm

Hartley, D. (2014, June 30). The Cultural Effects of Video Gaming -
 CertMag. Retrieved January 30, 2018, from http://certmag.com/
 the-cultural-effects-of-video-gaming/

Johnson, R. (2013). Hiding in Plain Sight: Reproducing Masculine
 Culture at a Video Game Studio. Communication, Culture &
 Critique, 7(4), 578-594. doi:10.1111/cccr.12023

Martins, N et al. (2009). "A Content Analyse of Female Body Imagery in
 Video Games." Sex Roles 56, 141–148.

Milford, T. (2016, September 29). Inequality in Gaming. Retrieved
 January 28, 2018, from http://www.equalityproject.ca/blog/
 inequality-in-gaming/

Taylor, R. (2015, April 14). Video Games' Place in American Culture.
 Retrieved January 29, 2018, from https://www.huffingtonpost.
 com/rich-taylor/video-games-place-in-american-
 culture_b_7063838.html

The SESHAT Volume 5 is an anthology assembled under the supervision of Drs. Austin and Catherine Mardon. This work will be published by the Golden Meteorite Press and promoted to different platforms such as Lulu, Google Scholar, and PubMed under the Antarctic Institute of Canada (AIC) Charity.

The SESHAT Volume 5 is not a double-blind peer-reviewed journal as most journals; however, all articles are peer-reviewed thoroughly by experienced premedical and graduate students, and Dr. Mardon. The articles accepted in this paper are authored by skilled Article Writers of the Antarctic Institute of Canada. This anthology serves to appreciate and showcase youth scholarly research in the fields of COVID-19, immigration and socioeconomic aspects of daily living to name a few.

There are no conflicts of interests to declare.

Special Thanks to the Editors, Daivat Bhavsar and Ehimen Ogadu, and the Graphic Designers Susie Woo and Amna Zia, for their relentless efforts in assembling the SESHAT Volume 5.

www.ingramcontent.com/pod-product-compliance
Lightning Source LLC
Chambersburg PA
CBHW031814190326
41518CB00006B/332